Representation of Multiple-Valued Logic Functions

Synthesis Lectures on Digital Circuits and Systems

Editor
Mitchell A. Thornton, *Southern Methodist University*

The Synthesis Lectures on Digital Circuits and Systems series is comprised of 50- to 100-page books targeted for audience members with a wide-ranging background. The Lectures include topics that are of interest to students, professionals, and researchers in the area of design and analysis of digital circuits and systems. Each Lecture is self-contained and focuses on the background information required to understand the subject matter and practical case studies that illustrate applications. The format of a Lecture is structured such that each will be devoted to a specific topic in digital circuits and systems rather than a larger overview of several topics such as that found in a comprehensive handbook. The Lectures cover both well-established areas as well as newly developed or emerging material in digital circuits and systems design and analysis.

Representation of Multiple-Valued Logic Functions
Radomir S. Stanković, Jaakko T. Astola, and Claudio Moraga
2012

Arduino Microcontroller: Processing for Everyone! Second Edition
Steven F. Barrett
2012

Advanced Circuit Simulation Using Multisim Workbench
David Báez-López, Félix E. Guerrero-Castro, and Ofelia Delfina Cervantes-Villagómez
2012

Circuit Analysis with Multisim
David Báez-López and Félix E. Guerrero-Castro
2011

Microcontroller Programming and Interfacing Texas Instruments MSP430, Part I
Steven F. Barrett and Daniel J. Pack
2011

Microcontroller Programming and Interfacing Texas Instruments MSP430, Part II
Steven F. Barrett and Daniel J. Pack
2011

Pragmatic Electrical Engineering: Systems and Instruments
William Eccles
2011

Pragmatic Electrical Engineering: Fundamentals
William Eccles
2011

Introduction to Embedded Systems: Using ANSI C and the Arduino Development Environment
David J. Russell
2010

Arduino Microcontroller: Processing for Everyone! Part II
Steven F. Barrett
2010

Arduino Microcontroller: Processing for Everyone! Part I
Steven F. Barrett
2010

Digital System Verification: A Combined Formal Methods and Simulation Framework
Lun Li and Mitchell A. Thornton
2010

Progress in Applications of Boolean Functions
Tsutomu Sasao and Jon T. Butler
2009

Embedded Systems Design with the Atmel AVR Microcontroller: Part II
Steven F. Barrett
2009

Embedded Systems Design with the Atmel AVR Microcontroller: Part I
Steven F. Barrett
2009

Embedded Systems Interfacing for Engineers using the Freescale HCS08 Microcontroller II: Digital and Analog Hardware Interfacing
Douglas H. Summerville
2009

Designing Asynchronous Circuits using NULL Convention Logic (NCL)
Scott C. Smith and JiaDi
2009

PSpice for Digital Signal Processing
Paul Tobin
2007

PSpice for Analog Communications Engineering
Paul Tobin
2007

PSpice for Digital Communications Engineering
Paul Tobin
2007

PSpice for Circuit Theory and Electronic Devices
Paul Tobin
2007

Pragmatic Circuits: DC and Time Domain
William J. Eccles
2006

Pragmatic Circuits: Frequency Domain
William J. Eccles
2006

Pragmatic Circuits: Signals and Filters
William J. Eccles
2006

High-Speed Digital System Design
Justin Davis
2006

Introduction to Logic Synthesis using Verilog HDL
Robert B.Reese and Mitchell A.Thornton
2006

Microcontrollers Fundamentals for Engineers and Scientists
Steven F. Barrett and Daniel J. Pack
2006

Copyright © 2012 by Morgan & Claypool

Representation of Multiple-Valued Logic Functions
Radomir S. Stanković, Jaakko T. Astola, and Claudio Moraga

ISBN: 978-3-031-79851-1 paperback
ISBN: 978-3-031-79852-8 ebook

DOI 10.1007/978-3-031-79852-8

A Publication in the Springer series
SYNTHESIS LECTURES ON DIGITAL CIRCUITS AND SYSTEMS

Lecture #37
Series Editor: Mitchell A. Thornton, *Southern Methodist University*
Series ISSN
Synthesis Lectures on Digital Circuits and Systems
Print 1932-3166   Electronic 1932-3174

Representation of Multiple-Valued Logic Functions

Radomir S. Stanković
University of Niš, Serbia

Jaakko T. Astola
Tampere University of Technology, Finland

Claudio Moraga
European Centre for Soft Computing, Spain
Technical University of Dortmund, Germany

SYNTHESIS LECTURES ON DIGITAL CIRCUITS AND SYSTEMS #37

ABSTRACT

Compared to binary switching functions, the multiple-valued functions (MV) offer more compact representations of the information content of signals modeled by logic functions and, therefore, their use fits very well in the general settings of data compression attempts and approaches. The first task in dealing with such signals is to provide mathematical methods for their representation in a way that will make their application in practice feasible.

Representation of Multiple-Valued Logic Functions is aimed at providing an accessible introduction to these mathematical techniques that are necessary for application of related implementation methods and tools.

This book presents in a uniform way different representations of multiple-valued logic functions, including functional expressions, spectral representations on finite Abelian groups, and their graphical counterparts (various related decision diagrams).

Three-valued, or ternary functions, are traditionally used as the first extension from the binary case. They have a good feature that the ratio between the number of bits and the number of different values that can be encoded with the specified number of bits is favourable for ternary functions.

Four-valued functions, also called quaternary functions, are particularly attractive, since in practical realization within today prevalent binary circuits environment, they may be easy coded by binary values and realized with two-stable state circuits. At the same time, there is much more considerable advent in design of four-valued logic circuits than for other p-valued functions.

Therefore, this book is written using a hands-on approach such that after introducing the general and necessarily abstract background theory, the presentation is based on a large number of examples for ternary and quaternary functions that should provide an intuitive understanding of various representation methods and the interconnections among them.

KEYWORDS

multiple-valued logic functions, Reed-Muller expressions, Reed-Muller-Fourier expressions, Galois field expressions, spectral expressions, decision diagrams

Contents

Acknowledgments

This work was supported in part by the Academy of Finland, Finnish Center of Excellence Programme, Grant No. 213462, and by the Foundation for the Advancement of Soft Computing, Mieres, Asturias, Spain.

Radomir S. Stanković, Jaakko T. Astola, and Claudio Moraga
May 2012

CHAPTER 1

Multiple-Valued Logic Functions

Discrete functions are usually defined as a mapping

$$f : \times_{i=1}^{n} S_i \rightarrow L, \tag{1.1}$$

where $S_i, i = 1, \ldots, n$, and L are finite non-empty sets of not necessarily equal cardinalities (in the discrete and finite case, the number of elements) $|S_i|$ and $|L|$, respectively, and $\times$ denotes the direct (Cartesian) product of sets. The sets $S = \times_{i=1}^{n} S_i$ and L are called the *domain* and the *range* for f, respectively.

In many cases, elements of the sets S_i and L in a fixed order are identified with the non-negative integers smaller or equal to $|S_i|$ and $|L|$, respectively, or with elements of an algebraic structure. This latter approach will be discussed in the context of optimization of function representations.

Many systems have several outputs and, therefore, for their modeling multi-output functions are used. In this case, the range of functions in (1.1) is a direct product of sets $L_i, i = 1, \ldots, k$, where k is the number of outputs of the system. Each output take values in a set L_i. Although these sets could be of different cardinalities, in practice it is usually assumed that all L_i are the same and, therefore, the range of a multiple-output function is taken as L^k.

Binary-valued functions and multiple-valued logic functions are two classes of discrete functions of particular interest in this book, with multiple-valued functions mostly, but not exclusively, viewed as generalizations of binary-valued functions. In these cases, $S_i = L = \{0, 1\}$, and $S_i = L = \{0, 1, \ldots, p - 1\}$, for all i, for binary and multiple-valued (p-valued) functions, respectively. For multiple-valued functions, the cases $p = 3$ and $p = 4$ are most often encountered in practice for pragmatic reasons. Ternary functions ($p = 3$) are most compact in the sense of the number of data which can be encoded with ternary sequences of the given length, while quaternary functions ($p = 4$) are convenient due to simple encoding of four values by binary sequences and then implementation with two-stable state circuits.

Binary-valued functions are often called Boolean functions, referring to the work by George J. Boole 1847 [18], [19], or *switching* functions, referring to the former means of their realizations by two-position relay networks or, alternatively, by two-states contact networks. In this book, we will mainly use the term switching for binary-valued functions. Although multiple-valued functions can be related to multi-positional switches, this term will not be applied to them.

Table 1.1: Truth-table for $xor5$							
$x_1x_2x_3x_4$	f	$x_1x_2x_3x_4$	f	$x_1x_2x_3x_4$	f	$x_1x_2x_3x_4$	f
00000	0	01000	1	10000	1	11000	0
00001	1	01001	0	10001	0	11001	1
00010	1	01010	0	10010	0	11010	1
00011	0	01011	1	10011	1	11011	0
00100	1	01100	0	10100	0	11100	1
00101	0	01101	1	10101	1	11101	0
00110	0	01110	1	10110	1	11110	0
00111	1	01111	0	10111	0	11111	1

1.1   TABULAR REPRESENTATIONS

Since discrete functions are mappings between finite sets, a straightforward way to specify a discrete function f is to enumerate its values at all the points of the domain of f. This can be done in a tabular form specifying the output for each combination of inputs, which can be simply represented as the vector of function values when the order of assignments for the variables is specified, or in some similar ways discussed below.

In general, the term *tabular* means enumeration of function values at all the points in the domain, or on certain subsets of these points, that are usually called *fields* or *arrays*. These subsets are selected depending on particular values a function can take in the range. It is assumed that at other points the functions take values that remain in the range. For instance, to specify a completely defined binary-valued function, it is sufficient to enumerate points where it takes the value 0 or 1. It is assumed that, in other points, the function takes the value opposite to the selected value.

In practice, there may be assignments of values for input variables that are highly improbable to appear as inputs in a function f, or the cases when f can take any value in the range for certain inputs. In these cases, f is called an *incompletely specified* function, and unspecified values are usually called *don't cares* and denoted by $-$ or $*$.

In the case of incompletely specified switching functions, it is necessary to specify two arrays, out of three possible arrays, for the values 1, 0, and the unspecified value $-$. The method of specifying binary functions by arrays is directly generalized to multiple-valued functions, and further to different classes of discrete functions. The term *in a similar way* used above, refers to such reduced representations by fields or arrays and some other related concepts that will be discussed and illustrated by the examples below.

Example 1 *Table 1.1 defines a binary-valued function of five variables which calculates the sum of the inputs modulo 2, and is known as $xor5$. The columns labeled by variables specify all the points of the domain of definition (the set of binary five-tuples). The columns labeled by f enumerates values of $xor5$ at the corresponding points.*

*If we fix an ordering among points in the domain as, for example, the lexicographic ordering used in Table 1.1, it is sufficient to state this ordering and the function values for a given f as a vector **F**, which is often called the truth-vector for f.*

The truth-vector of xor5 is

$$\mathbf{F} = [0, 1, 1, 0, 1, 0, 0, 1, 1, 0, 0, 1, 0, 1, 1, 0, 1, 0, 0, 1, 0, 1, 1, 0, 0, 1, 1, 0, 1, 0, 0, 1]^{T}.$$

Since the function xor5 take two values, 0 and 1, it is sufficient to specify points (or assignments of variables) for any of them, and assume that in the remaining points the function takes the opposite value.

For instance, the function xor5 is completely specified by the 1-field or 1-array, i.e., by the set of points where xor5 has the value 1.

		.i5	
		.o1	
		11111	1
		01110	1
		10110	1
1-array for xor5		00111	1
00001	10000	11010	1
00010	10011	01011	1
00100	10101	10011	1
00111	10110	00010	1
01000	11001	11100	1
01011	11010	01101	1
01101	11100	10101	1
01110	11111	00100	1
		11001	1
		01000	1
		10000	1
		00001	1
		.e	

Such a specification is also called PLA specification of xor5. The term PLA is devised by referring to the realizations with Programmable Logic Arrays. *In practice, it is usual to specify the number of inputs and outputs, and indicate the end of the file, as is done by symbols (.i), (.o), and (.e) in the PLA specification of xor5.*

Alternatively, we can specify xor5 by the 0-array. This selection is important in the case of functions with different distribution of function values. For instance, if a switching function takes the value 1 at many points (more than $2^{n/2}$), then it is economic in terms of space to represent it by the 0-array.

Table 1.2: PLA specification of the MCNC benchmark function *con* 1	
.i 7	
.o 2	
.p 9	
-1- -1- -	10
1-11- - -	10
-001- - -	10
01- - -1-	10
-0- -0- -	01
1- - -0- -	01
0- - - - -0	01

1.2   CUBES

Instead of writing the complete assignments of values for inputs in an array (PLA format), it is sufficient to show their contracted representation obtained by extracting common parts under a convention that a bar (-) can replace either 0 or 1. This contracted expression covering few points is called a cube.

Example 2 *If a switching function of four variable takes the same value at the points* 0010 *and* 0011, *we can represent both points by a cube* 001−, *were the bar (−) stands for either* 0 *and* 1.

Table 1.2 *specifies the MCNC (Microelectronics Center of North Carolina) benchmark function* con1. *This function has* 7 *inputs,* 2 *outputs, and it is specified by* 9 *cubes. This is an example of tabular representations of multi-output functions.*

The number of unspecified values in a cube determines the *order* of the cube. In some publications, the order of a cube of n variables with k unspecified bits is defined as $n - k$.

Tabular representations and their reduced forms (cubes and PLAs) for switching functions can be extended in a straightforward way to the representations of multiple-valued functions.

Example 3 *Table 1.3 shows the truth-table of a function* $f : \{0, 1, 2, \}^2 \rightarrow \{0, 1, 2\}$, *that is the sum of two ternary variables modulo* 3.

The same function can be alternatively specified by either the 0-*array and* 1-*array, or* 0-*array and* 2-*array, or* 1-*array and* 2-*array.*

0-array	1-array		0-array	2-array		1-array	2-array
00	01		00	02		01	02
12	10		12	11		10	11
21	22		21	20		22	20

Table 1.3: Truth-table of a ternary function in Example 3

$x_1 x_2$	f
00	0
01	1
02	2
10	1
11	2
12	0
20	2
21	0
22	1

Multiple-valued input binary-valued output functions

$$f : \{0, 1, \ldots, p - 1\}^n \to \{0, 1\}$$

are a particular class of multiple-valued functions that can be used in representation of switching functions and solving certain optimization problems related to these functions [167]. An example of such applications is reduction of the area in PLAs with address decoders, as proposed in [164]. The method uses the feature that a switching function with n inputs and k outputs, can be represented by a single-output function of n binary variables and the k-valued $(n + 1)$-st variable.

Definition 1 *(Characteristic functions)*
If $f = (f_1, \ldots, f_k)$, where $f_j = f_j(x_1, \ldots, x_n)$, $j = 0, \ldots, k - 1$, then the characteristic function for f is $F : \{0, 1\}^n \times \{0, 1, \ldots, k - 1\}$ defined by $F(a_1, \ldots, a_n, j) = f_j(a_1, \ldots, a_n)$ for $(a_1, \ldots, a_n) \in \{0, 1\}^n$ and $j \in \{0, 1, \ldots, k - 1\}$.

The background idea for introducing the characteristic functions defined in this way is that the minimization of binary input multiple-output binary functions, can be done through the minimization of single output binary-valued functions of multiple-valued inputs. Further applications of multiple-valued input binary-valued output functions are synthesis with FPGAs [169].

Example 4 *Table 1.4 shows a three-input three-output function describing the functioning of an adder. The outputs are the sum s_i, the carry c_i and its complement $\overline{c}_i$. The so-called characteristic function f_c for this system of switching functions has three binary variables x_1, x_2, x_3 and a three-valued variable x_4 due to three outputs. Table 1.5 shows f_c.*

Table 1.4: Three-bit adder

$x_i y_i c_{i-1}$	s_i	c_i	$\overline{c}_i$
000	0	0	1
001	1	0	1
010	1	0	1
011	0	1	0
100	1	0	1
101	0	1	0
110	0	1	0
111	1	1	0

Table 1.5: Characteristic function for an adder

$x_1 x_2 x_3 x_4$	f_c	$x_1 x_2 x_3 x_4$	f_c
0000	0	1000	1
0001	0	1001	0
0002	1	1002	1
0010	1	1010	0
0011	0	1011	1
0012	1	1012	0
0100	1	1100	0
0101	0	1101	1
0102	1	1102	0
0110	0	1110	1
0111	1	1111	1
0112	0	1112	0

Table 1.6: Different domains for variables in a function defined in 16 points

n	$x_1x_2x_3x_4$	f	x_1x_2	f	x_1x_2	f
0.	0000	$f(0000)$	00	$f(00)$	00	$f(00)$
1.	0001	$f(0001)$	01	$f(01)$	01	$f(01)$
2.	0010	$f(0010)$	02	$f(02)$	02	$f(02)$
3.	0011	$f(0011)$	03	$f(03)$	03	$f(03)$
4.	0100	$f(0100)$	10	$f(10)$	04	$f(04)$
5.	0101	$f(0101)$	11	$f(11)$	05	$f(05)$
6.	0110	$f(0110)$	12	$f(12)$	06	$f(06)$
7.	0111	$f(0111)$	13	$f(13)$	07	$f(07)$
8.	1000	$f(1000)$	20	$f(20)$	10	$f(10)$
9.	1001	$f(1001)$	21	$f(21)$	11	$f(11)$
10.	1010	$f(1010)$	22	$f(22)$	12	$f(12)$
11.	1011	$f(1011)$	23	$f(23)$	13	$f(13)$
12.	1100	$f(1100)$	30	$f(30)$	14	$f(14)$
13.	1101	$f(1101)$	31	$f(31)$	15	$f(15)$
14.	1110	$f(1110)$	32	$f(32)$	16	$f(16)$
15.	1111	$f(1111)$	33	$f(33)$	17	$f(17)$

1.3 ENCODING OF VARIABLES

A function defined in a given number of points can be viewed as a function of different number of variables taking values in different sets. This can be also viewed as different encoding of disjoint subsets in the initial set of variables. Selection of the encoding that is the best suited to an application can be of importance in many cases. For instance, this approach was used in optimization of realizations with PLA [164], [167], synthesis by FPGAs [169], reduction of decision diagrams [188], [190], and some other applications [167].

Example 5 *Table 1.6 illustrates that a function defined in* 16 *points can be viewed as a function of 4 binary variables* $x_1, x_2, x_3, x_4 \in \{0, 1\}$, *two 4-valued variables* $x_1, x_2 \in \{0, 1, 2, 3\}$, *a binary variable* $x_1 \in \{0, 1\}$, *and an 8-valued variable* $x_2 \in \{0, 1, 2, 3, 4, 5, 6, 7\}$.

The encoding of variables can be interpreted as the change of the domain group for the given function and is used in the group-theoretic approach to the optimization problems in multiple-valued logic [188], [190], [191], [209], [214], [215].

1.4 OTHER REPRESENTATIONS

Tabular representations, as well as their reduced forms such as cubes, are often inefficient for functions of a large number of variables, i.e., functions defined in many points, due to their exponential complexity, especially, such representations are of a limited use in various manipulations and calculations with discrete functions. For this reason, various analytical representations as Sum-of-Products or Product-of-Sums expressions were investigated already by De Morgan in 1874 [34]. This subject was of a continuous interest in the past, and nowadays it has a renewed importance due to demands coming from features that are present in contemporary or can be expected in future technologies for realization of digital systems [22]. We provide a few references to the former and the recent work illustrating different attempts to define various functional expressions for discrete functions [5], [6], [7], [8], [14], [31], [40], [65], [66], [71], [77], [85], [98], [102], [105], [111], [145], [152], [184], [205], [250], [251], [254], [264].

Sum-of-Products expressions and their counterparts Product-of-Sums will be considered in Chapter 2. In the same chapter, we also discuss the Reed-Muller expressions that are particular functional expressions for binary-valued functions which can be viewed as a discrete analog of either Taylor series or Fourier series for functions on the real line $\mathcal{R}$. They are defined in terms of a particular set of basis functions defined as elementary products of binary variables. These are products of all possible combinations of n binary-valued variables. For a given function f, no identical products can appear in the Reed-Muller expression for f. These representations have attracted a considerable attention in the last two decades, which can be seen from the specialized workshops devoted to this subject, starting from 1993.

In Chapter 2, we further discuss extensions and generalizations of Reed-Muller expressions defined by preserving the same set of basis functions for the binary case or its straightforward generalizations for the multiple-valued case. We attempt to provide explanations of differences and motivation to introduce few different expressions that are considered in the literature.

These representations can be viewed as particular classes of spectral representations with respect to certain appropriately selected sets of basis functions that will be discussed in Chapter 3. The central subject of Chapter 4 is the graphical representation of multiple-valued functions by various decision diagrams. These diagrams are viewed as graphical counterparts of the corresponding functional expressions.

1.5   ALGEBRAIC STRUCTURES FOR MULTIPLE-VALUED FUNCTIONS

The switching (Boolean) functions are usually studied assuming that the underlying algebraic structures are the Boolean algebra $B(\{0, 1\}, \vee, \wedge)$ and the Boolean ring $B(\{0, 1\}, \oplus, \wedge, -)$, where the binary operations have the following meaning: $\vee$ is the logic OR, $\wedge$ is logic AND and $\oplus$ is the logic EXOR (addition modulo 2). The unary operation $-$ is the logic NOT (the logic complement of logic values 0 and 1). Recall that the transition from the Boolean algebra to the Boolean ring and vice versa is always possible due to the relationship between the used operations $x \oplus y = \overline{x}y \vee x\overline{y}$. Note that the set of all Boolean functions of a given number of variables also endorses the structure of the Boolean algebra or the Boolean ring with the corresponding Boolean operations applied componentwise.

In the multiple-valued case, the structure of a Boolean algebra is not sufficient with respect to the completeness; for instance, see [145]. A Boolean algebra can represent 2^{2^n} functions, while in the p-valued case, there are possible p^{p^n} functions; thus, just a portion of them can be represented by a Boolean algebra. In the case when p is a power of 2, the generalized Boolean algebras can be used [145].

Therefore, for multiple-valued functions, various algebras are defined in terms of different operations. A great variety of possible choices becomes clear if we recall that in the p-valued case, there are p^{p^2} possible binary operators and p^p unary operators.

When looking for the origins of the multiple-valued logics and related algebraic structures as their models, it should be mentioned that Aristotle, the father of the binary logic, pointed out problems in applying laws of binary logic to future events, as he elaborated in his discussion of the famous sea battle paradox (Aristotle, *De Interpretatione*, Chapter 9). It is, however, incorrect to say that he anticipated introducing a system of logic with more than two truth values.

In [162], as forerunners of multiple-valued logic, A. Salomaa mentioned Hugh MacColl and Charles Sanders Peirce. Salomaa pointed out that Peirce used to speak of logic of three dimensions and also trichotomic mathematics. These fragmentarily presented ideas can be found in an unpublished manuscript *Minute Logic* from 1902 by C.S. Peirce, reprinted in [149].

It is of course stated on p. 118 of [162] that "Actual discovery of many-valued logic was made independently by Lukasiewicz and Post about 1920." See, [108], [109], [150], [151]. This is a widely accepted point of view with the difference in the motivations for the work between Lukasiewicz and Post. It can be said that the work of Lukasiewicz was primarily motivated by philosophical problems, while Post aimed at mathematical extensions of the bivalent logic; for instance, see [29] and references therein.

Salomaa also mentions the work of Vasiliev in 1924 [242], and Edwin Guthrie in 1916 [73], although he remarked that the former did not develop his ideas any further.

Notice that the logic of Lukasiewicz was axiomatized by Moisil [124], [125], [126] for the cases ternary and quaternary logic. In further work (for instance, see a summary of it in [127]), Moisil introduced the n-valued Lukasiewicz algebra, however, as pointed by A. Rose, for $n \geq 5$, the Lukasiewicz implication cannot any more be defined on a Lukasiewicz algebra. Thus, n-valued Lukasiewicz algebras are models for another system of logic introduced by Moisil, and these algebras are now often called Lukasiewicsz-Moisil algebras [29] and the corresponding logic is called the Moisil logic.

After these introductory remarks, it might be summarized that the work in multiple-valued algebras can be classified as being oriented towards three directions:

1. philosophical work towards finding interpretations of concepts as truth, possibility, and necessity (like Lukasiewicz);

2. mathematical work aimed at providing as close as possible extensions of the Boolean structures assuring the completeness (like Post); and

3. engineering-oriented approaches trying to be based on operations easy implementable in circuits.

In this book, we are primarily interested in the second and third directions of research, viewing at the same time the second as attempts to provide algebraic foundations to philosophical work.

In the second direction of research, in many cases the used operations are derived from various interpretations of the corresponding Boolean operations. For instance, the logic AND is viewed as the minimum *min* and logic OR as the maximum *max*, while EXOR is considered as addition modulo p.

It is widely accepted that the first complete work towards a multi-valued algebra is due to Emil L. Post [150], who used two operators: the unary operator $(1 + x_i) \bmod p$ corresponding to the logic NOT and the binary operator maximum $max(x_i, x_j)$ corresponding to the logic OR. These two operators ensure the completeness, in the same way as the logic NOR does in the binary case; see also [151].

The operators *min* and *max* are combined by many authors with various variants of the unary operators defined as particular mappings of the set of input values $\{0, 1, \ldots, p - 1\}$ into a subset S of it. In general, the unary operators are mainly defined as analogs to either the negation operator or the literal operators, however, they can be defined also in other ways. Recall that there is a number of equivalences among different unary operators and they can be derived from each other. Examples of unary operators defined as a generalization of logic complement in the binary case are:

1. negation $\overline{x}_i = (p - 1) - x_i$;

2. successor $\vec{x}_i = (1 + x_i) \bmod p$; and

3. predecessor defined as $(x_i - 1) \bmod p$.

Examples of multiple-valued literals are:

1. The selection literal

$$x^S = \begin{cases} p - 1, & \text{if } x \in S, \\ 0, & \text{otherwise.} \end{cases}$$

2. The threshold literal

$$x^a = \begin{cases} p - 1, & \text{if } x \geq a, \\ 0, & \text{otherwise.} \end{cases}$$

3. The decisive literal

$$C_i(x) = \begin{cases} p - 1, & \text{if } x = i, \\ 0, & \text{otherwise.} \end{cases}$$

These literals defined in various ways were used to provide different axiomatizations of the multiple-valued algebras. For example, the decisive literal was used in this sense by G. Epstein in [41], [42], [43], [44], [45]. The work was motivated by the desire of throwing the maximum operator exploiting essential properties of the used unary operators.

In the same context, the work by D.L. Webb should be mentioned. The Webb algebra [246] is defined by introducing an analog of the logic NOR (the Sheffer stroke) operator for multiple-valued logic defined as

$$x_i | x_j = \begin{cases} (1 + x_i) \bmod p, & \text{if } x_i = x_j, \\ 0, & \text{otherwise.} \end{cases}$$

The particular case of the Webb algebra for $p = 3$ is considered in [113].

The p-valued algebras based upon the addition and multiplication modulo p were considered by many authors, see for instance the early works in this area [15], [16], [31], [152], [153], leading eventually to generalizations of the Reed-Muller expressions in binary-valued logic and various related forms as the Tamari expressions [225] and the Reed-Muller-Fourier expressions [184], [205].

The third direction of research is related with algebras for multiple-valued functions defined in terms of appropriately defined operations, the introduction of which is motivated by certain properties the operators and the related algebra may express or aimed at achieving some particular features in their circuit implementations, some of them directly driven by the targeted technology to implement multiple-valued functions in related circuitry. Examples of these implementation oriented algebras are presented in [9], [40], [104], [243]. For example, the algebra proposed in [104] introduces an operator directly corresponding to the ternary multiplexer since its action is described as

$$T(A, B, C, D) = \begin{cases} A, & \text{if } D = 0, \\ B, & \text{if } D = 1, \\ C, & \text{if } D = 2. \end{cases}$$

With the same motivation as in [40], to provide an algebraic framework for CMOS realizations of multiple-valued circuits, an algebra based on suitably defined threshold operators was introduced in [25], [26]. For a further discussion of various multiple-valued algebras, see, for instance [82], [119], [145], [159], [160].

In this book, we will mainly, although not exclusively, assume a signal processing oriented approach to the processing of logic signals [12], [82], and view multiple-valued functions as functions defined on finite groups taking values in either finite fields of the corresponding order or the field of complex numbers. We use the property that the set of all multiple-valued functions for the given p and n endorse the structure of the Hilbert space, which permits the usage of all efficient mathematical tools on these structures when representing multiple-valued functions. For definitions and properties of these algebraic structures we refer to classical books on mathematical analysis or possibly to publications more oriented towards engineering applications [11], [82], [99].

1.6   FUNCTIONS WITH VARIOUS PROPERTIES

If a function expresses some peculiar property, this property can be useful in certain applications, i.e., the function can be realized by a simpler network. The analysis of this network can also be simpler than in the general case [28], [163], [202], [203].

Therefore, classes of functions with particular properties are important in practice. In this book, we will use some such functions as examples. For this reason, in this chapter we define a couple of such examples aiming to show that many definitions for binary-valued functions can be extended to multiple-valued functions. At the same time, we should note that some other definitions, for instance various ways to define the logic negation, are peculiar for multiple-valued functions.

Definition 2 *A linear function is the constant 0 or the sum of some of the variables. The sum is defined in the underlined algebraic structure assumed for the study of the function considered. It could be the modulo p sum, as for instance the logic EXOR in the case of switching functions, the sum in the finite field, the sum in the field of complex numbers in the case of integer-valued functions, etc.*

Definition 3 *In the case of switching functions, an affine function is a linear function or the complement of a linear function. In the general case, an affine function is a vector-valued function of the form*

$$f(x_1, \ldots, x_n) = \mathbf{A}_1 \mathbf{x}_1 + \cdots + \mathbf{A}_n \mathbf{x}_n + \mathbf{b},$$

where $\mathbf{A}_i, i = 1, \ldots, n$ can be scalars of matrices, and $\mathbf{b}$ is a scalar or a vector. Thus, an affine transformation consists of a linear transformation defined by the matrices $\mathbf{A}_i$ followed by a translation defined by the vector $\mathbf{b}$. In the binary case, $\mathbf{A}_i$ and $\mathbf{b}$ may be 0 and 1 and the addition is the modulo 2 addition (EXOR).

In the case of switching functions, the complement of a function f is defined as $f \oplus 1$. In the general case, this is expressed as the addition of f and the vector $\mathbf{b}$. Notice that in the case of multiple-valued functions, the complement of a logic value x can be defined in different ways depending on the underlying algebraic structure. Most often, the complement of a p-valued variable x is defined as $\overline{x} = p - x$, where subtraction is the arithmetic subtraction in the set of integers.

Definition 4 *A symmetric function is a function that remains unchanged by any permutation of its variables.*

Example 6 *The function $x \, or \, 5$ in Example 1 is symmetric. The ternary function in Example 3 is symmetric.*

For more discussions about symmetric multiple-valued functions, see [23], [24], [46], [47], [106], [115], [141], [143], [171], [198], [217], [224], [238], [252].

Definition 5 *The weight w_f of a function f is the number of non-zero values in the function vector of f. In the binary case, the weight is the number of 1 values in the truth-vector.*

Definition 6 *An n-variable switching function f is balanced iff its weight is 2^{n-1}. Thus, a balanced switching function has the same number of 0 and 1 values. In the same way, a p-valued function is balanced if the number of all possible different function values is the same.*

Table 1.7: Functions f_1 and f_2 in Example 8

x_1x_2	f_1	f_2	x_1x_2	f_1	f_2
00	0	0	20	2	2
01	1	1	21	3	3
02	2	2	22	0	0
03	3	3	23	1	1
10	1	1	30	3	3
11	2	0	31	0	2
12	3	3	32	1	1
13	0	2	33	2	0

Example 7 *The switching function x or 5 in Example 1 is balanced, since there is the equal number of 0 and 1 values in the truth-vector for f. The ternary function in Example 3 is balanced, since there is the same number of 0, 1, and 2 values in the function vector in Table 1.3.*

Definition 7 *The Hamming distance of two functions f_1 and f_2 is the number of positions in the function vectors for $\mathbf{F}_1$ and $\mathbf{F}_2$ where f_1 and f_2 differ. In other words, the Hamming distance is the minimum number of substitutions that converts two functions each to other.*

Example 8 *Table 1.7 shows two functions*

$$\begin{aligned} f_1 &= x_1 \oplus_4 x_2, \quad \text{addition modulo } 4, \\ f_2 &= x_1 + x_2, \quad \text{addition in } GF(4). \end{aligned}$$

The Hamming distance of these functions is 4, since their values differ at points $(1, 1)$, $(1, 3)$, $(3, 1)$, and $(3, 3)$.

CHAPTER 2

Functional Expressions for Multiple-Valued Functions

In this chapter, we present functional expressions for multiple-valued logic functions that can be viewed as generalizations of the corresponding representations for binary logic functions.

2.1 FUNCTIONAL EXPRESSIONS

Functional expressions for discrete functions can be viewed as formulae specifying the behavior of functions. In other words, they describe uniquely the mapping between the domain (the set where the variables take values) and the range (the set where the function takes its values), defining the function considered. A functional expression consists of symbols for variables, symbols for functions, and symbols for operations over variables and functions. To determine the meaning of symbols for operations, some algebraic structures, not necessarily identical, are imposed on the domain and the range. In a functional expression, variables are organized into terms by operations over variables (product terms when variables are connected by the operation corresponding to the multiplication in the assumed underlying algebraic structure) related by operations over terms. A possible hierarchy of operations over variables and terms is often specified by brackets. Depending on the algebraic structures (equivalently, operations) and restrictions possibly imposed on product terms, various functional expressions are defined. From a given functional expression, some other expressions can be derived by using postulates and theorems in the assumed algebraic structures. For instance, in probably the most widely known functional expressions for binary logic functions, the disjunctive normal form (DNF), the underlying algebraic structure is the two-element Boolean algebra ($\{0, 1\}, \vee, \wedge, -$) and variables are organized into product terms under logic AND ($\wedge$), while the product terms are connected by the logic OR ($\vee$). A restriction imposed to product terms is that a variable can have a single appearance in a term whatever subjected to logic negation or not. This restriction comes from the Boolean postulate $x \wedge x = x$. This class of functional expressions is also called the Sum-of-Product (SOP) expressions or AND-OR expressions. By using De Morgan laws, the disjunctive normal form can be converted in the conjunctive normal form (CNF) or Product-of-Sum (POS) expressions which are another example of AND-OR expressions.

The transition from the Boolean algebra to the Boolean ring $B_2 = (\{0, 1\}, \oplus, \wedge)$ and vice versa is always possible, which leads to another class of functional expressions for binary logic functions called Zhegalkin polynomials [260], [261], or Reed-Muller expressions [142], [158]. In these expressions, variables are again organized in products under logic AND, however, under the restriction that products consist of all possible disjoint subsets of variables. In other words, products are obtained as logic AND over elements of entries of the power set for the set of variables $x_1, \ldots, x_n$. The product terms are related by modulo 2 addition, also called logic EXOR ($\oplus$). Due to that, these functional expressions can be viewed as expressions assuming the Galois field of order 2, $GF(2)$, as the underlying algebraic structure. These expressions belong to the class of AND-EXOR expressions. For other expressions in the same class, see for instance [11], [166], [167], [195].

Recall that besides algebraic structures imposed to the domain and the range of multiple-valued functions, the set of all p-valued functions of a given number of variables n might also be enriched with appropriately defined operations to exhibit certain algebraic structure. Dealing with such structures and their study proved useful in the study of logic functions and their applications.

A brief review of certain possible alternatives for underlying algebraic structures for the set of all n-variable functions, in the case of binary functions, can be a good basis to discuss generalizations to the multiple-valued functions.

For instance, the set B_{2^n} of all n-variable binary switching functions also endorse the structure of a Boolean ring, or alternatively a Boolean algebra, under the corresponding logic operations AND, OR, and EXOR, but this time taken componentwise over 2^n vectors. In that setting, the coefficients in the Reed-Muller expressions can be expressed in terms of the Boolean difference [254] that is accepted as an operator over the set B_{2^n} of all n-variable switching functions expressing properties of a differential operator [228].

When considered as elements of a discrete function space, the Reed-Muller expressions for binary functions are a finite dyadic field counterpart of the Taylor series expansions of real variable functions [67], [69], [183]. The values of the coefficients in these expressions, i.e., the Reed-Muller coefficients, are again viewed as values of Boolean differences of the order corresponding to the number of variables in the products, in the similar way as the coefficients in Taylor expansions are values of Newton-Leibniz derivatives [251], [254]. Therefore, this approach to derive Reed-Muller expressions can be called the polynomial oriented approach [213].

Assuming the Galois field $GF(2)$ for both the domain and the range of switching functions, these functions can be considered as mappings $f : (GF(2))^n \to GF(2)$ and their polynomial representations can be interpreted as Galois field expressions. However, Reed-Muller coefficients can be interpreted as the coefficients of the so-called Fourier-Galois transform acting in the set of all n-variable switching functions [66], [98], i.e., as coefficients of a Fourier-like transform over $GF(2)$. This approach can be denoted as the Fourier-oriented approach. It proved very useful in calculation of Reed-Muller coefficients through FFT-like algorithms [17]. The computation method was extended to decision diagrams representation of switching functions in [30].

The set of all Boolean functions is viewed as the support set of the finite dyadic field. If this field is enriched by the convolutionwise Gibbs multiplication, the Gibbs algebra is devised [68]. When the Reed-Muller expressions are considered in this algebra, they are viewed as Fourier-like expressions. The term *Instant Fourier transform* was used to emphasize the resemblance to the properties of the Fourier series for real-valued functions of real variables [68]. A generalization of this algebra to multiple-valued functions is used to define another class of functional expressions called the Reed-Muller-Fourier expressions which express a stronger similarity with the Fourier-series expressions than the Reed-Muller expressions.

Assume the complex field or the field of rational numbers as the range of switching functions permits derivation of the so-called arithmetic transform representations of switching functions that are polynomial representations of the form directly analogous to the Reed-Muller expressions, but with integer-valued coefficients [78], [144]. For the extension of arithmetic transform to multiple-valued case, see [112], [156], [157], [206], [250] and discussions below.

2.2   GENERALIZATIONS TO MULTIPLE-VALUED FUNCTIONS

For the considerations in this chapter in order to easier distinguish different approaches in defining various functional expressions for multiple-valued functions, the algebraic structures assumed for the domain and the range of logic functions will be conditionally called the variable-related and function-related algebraic structures, respectively.

Unlike the variety of different possible approaches to derive the polynomial representations of binary logic functions, there are some considerable limitations in extending the theory of polynomial representations to multiple-valued functions.

1. Regarding the variable-related algebraic structures, the approach based upon the Boolean algebra fails for the problem of completeness [85], [145].

2. Ring representations based upon the modulo p addition and multiplication can be derived, but the corresponding transform matrices for the calculation of polynomial coefficients have proportionally more non-zero elements than in the binary case and, thus, are often inefficient in practical applications.

3. Galois field representations can be derived, see, for example [145], but have some disadvantages especially for the non-prime p. In particular, the complexity of realization of the corresponding addition and multiplication is considerably greater than for prime p, since for p-prime the field operations reduce to the modulo p addition and multiplication.

   Together with that, the calculation methods for the determination of the coefficients of these representations that are based upon the Newton's divided differences are rather complicated; see discussions in [145]. Some improvements of the procedure for $p = 2, 3$, and 4 are given in [263] by taking the advantages from certain peculiar properties of these fields. Note that the Newton's divided difference method does not allow the optimization of the polynomial representations since it does not permit the use of the negative literals for switching variables[1].

---

[1]The use of negative literals for switching variables can be interpreted as a reordering of elements in the domain of switching functions, but the Newton's divided differences method produces the polynomial representations of the same form irrespective to the used ordering.

4. The variable-oriented algebraic structures approaches based upon the Walsh transform can be extended to multiple-valued functions through the Vilenkin-Chrestenson transform [129], [134], but involves the complex number arithmetic in dealing with multiple-valued functions [98], [129], which can be considered as a disadvantage of the approach. The advantage is that there are fast algorithms for the determination of the Vilenkin-Chrestenson transform coefficients based upon the FFT-like algorithms [98] and multiple-place decision diagrams [206], the second permitting processing of large functions.

The spectral techniques approach enables the derivation of FFT-like algorithms for computation of the coefficients of Galois field representations through a suitable modification of the Fast Vilenkin-Chrestenson transform; for example, see [77]. Moreover, the approach permits the consideration of p^n different polarity Galois field representations whose coefficients can be calculated through suitable reorderings of the weighting coefficients in the fast flow-graphs. These different polarities correspond to some particular permutations of p-valued switching variables. The consideration of other permutations among $(p!)^n$ possible permutations does not affect the number of non-zero coefficients in the Galois field expressions and, therefore, is not efficient in applications.

5. The extension of the function related algebraic structure approach based upon the algebra of all p-valued functions for a given n under the corresponding extension of the Gibbs multiplication was studied for any prime p in [184] and for non-prime p in [205]. The coefficients of the polynomial representations are interpreted as the coefficients of a Fourier-like transform for multiple-valued functions [184]. The partial case for $p = 3$ was discussed in [209].

The transfer back to the variable-oriented algebraic structures through some p-adic multiplication and exponentiation based again upon the Gibbs multiplication was suggested in [205]. This approach resembles the relationship between the two-element Boolean algebra or the Boolean ring assumed for the domain and the range of binary switching functions and the corresponding structures over the set of all n-variable binary logic functions. These representations are denoted as the Reed-Muller-Fourier representations due to properties of the corresponding transform matrices. Moreover, thanks to some of their properties, the FFT-like algorithm and multiple-place decision diagrams methods for calculation of Reed-Muller-Fourier coefficients are proposed [222]. Therefore, greater computation efficiency [205] compared to the Newton's divided differences methods, the existence of the fast calculation algorithms, and the possibility to consider the optimization procedures based upon the choice among the total of $(p!)^n$ compared to p^n different polarity Galois field representations [59], [60], [77], are reasons to give some advantage to that approach in extending the theory of polynomial representations to multiple-valued functions. With this motivation, the Gibbs algebra-based approach was extended to multiple-valued functions [184].

In the following sections, we will consider several different functional expressions for multiple-valued functions and then point out some possibilities for their optimization in the sense of reducing the number of non-zero coefficients.

2.3   SUM-OF-PRODUCT EXPRESSIONS

In this section, we present a generalization of SOP expressions to the multiple-valued case.

Consider a ternary function $f(x)$ defined by the function vector $\mathbf{F} = [1, 2, 0]^T$. It can be written as the sum of three vectors with a single non-zero element as

$$
\mathbf{F} = \begin{bmatrix} 1 \\ 2 \\ 0 \end{bmatrix} = 1 \cdot \begin{bmatrix} 1 \\ 0 \\ 0 \end{bmatrix} \oplus 2 \cdot \begin{bmatrix} 0 \\ 1 \\ 0 \end{bmatrix} \oplus 0 \cdot \begin{bmatrix} 0 \\ 0 \\ 1 \end{bmatrix},
$$
$$
= 1 \cdot J_0(x) \oplus 2 \cdot J_1(x) \oplus 0 \cdot J_2(x),
$$

where

$$
J_0(x) = \begin{bmatrix} 1 \\ 0 \\ 0 \end{bmatrix}, \quad J_1(x) = \begin{bmatrix} 0 \\ 1 \\ 0 \end{bmatrix}, \quad J_2(x) = \begin{bmatrix} 0 \\ 0 \\ 1 \end{bmatrix}.
$$

Thus, we define

$$
J_i(x) = \begin{cases} 1, & \text{if } i = x, \\ 0, & \text{otherwise}, \end{cases}
$$

which can be viewed as a generalization of the notion of the positive and the negative literals $x = \begin{bmatrix} 0 \\ 1 \end{bmatrix}$ and $\bar{x} = \begin{bmatrix} 1 \\ 0 \end{bmatrix}$ in binary logic.

The same principle can be extended to functions of an arbitrary number of variables n and an arbitrary value of p.

Definition 8 *(Characteristic functions)*
For a multiple-valued variable x_j taking values in the set $\{0, 1, \ldots, p-1\}$, $j = 0, \ldots, m-1$, the characteristic functions $J_i(x_j)$, $i = 0, 1, \ldots, p-1$, are defined as $J_i(x_j) = 1$ for $x_j = i$, and $J_i(x_j) = 0$ for $x_j \neq i$.

The functions $J_i(x_j)$ are viewed as functions of n variables, however, essentially dependent on the variable x_j. In the matrix notation, these functions are written as vectors of length p^n and can be multiplied by the Hadamard product to derive a basis for the representation of functions of n variables.

Example 9 *For $p = 3$ and $n = 2$, the characteristic functions $J_i(x_j)$ are given in Table 2.1.*

Table 2.1: Characteristic functions for $p = 3$ and $n = 2$

$x_1 x_2$	$J_0(x_1)$	$J_1(x_1)$	$J_2(x_1)$	$J_0(x_2)$	$J_1(x_2)$	$J_2(x_2)$
00	1	0	0	1	0	0
01	1	0	0	0	1	0
02	1	0	0	0	0	1
10	0	1	0	1	0	0
11	0	1	0	0	1	0
12	0	1	0	0	0	1
20	0	0	1	1	0	0
21	0	0	1	0	1	0
22	0	0	1	0	0	1

Definition 9 *Every ternary function* $f(x_1, \ldots, x_n)$ *can be expressed as*

$$f(x_1, \ldots, x_n) = \sum_{a_1, a_2, \ldots, a_n \in Q} f(a_1, a_2, \ldots, a_n) J_{a_1}(x_1) J_{a_2}(x_2) \cdots J_{a_n}(x_n), \qquad (2.1)$$

where Q is the set of all ternary n–tuples and summation is modulo 3, and characteristic functions are multiplied by the Hadamard product.

Example 10 *Any ternary function of three variables can be written as*

$$
\begin{aligned}
f(x_1, x_2) = \; & f(00) J_0(x_1) J_0(x_2) \oplus f(01) J_0(x_1) J_1(x_2) \oplus f(02) J_0(x_1) J_2(x_2) \\
& \oplus f(10) J_1(x_1) J_0(x_2) \oplus f(11) J_1(x_1) J_1(x_2) \oplus f(12) J_1(x_1) J_2(x_2) \\
& \oplus f(20) J_2(x_1) J_0(x_2) \oplus f(21) J_2(x_1) J_1(x_2) \oplus f(22) J_2(x_1) J_2(x_2).
\end{aligned}
$$

Consider, for instance, the ternary function $f(x_1, x_2) = x_1 \oplus x_2$, *whose function vector is* $\mathbf{F} = [0, 1, 2, 1, 2, 0, 2, 0, 1]^T$. *This function can be written as*

$$\mathbf{F} = \begin{bmatrix} 0 \\ 1 \\ 2 \\ 1 \\ 2 \\ 0 \\ 2 \\ 0 \\ 1 \end{bmatrix} = \begin{bmatrix} 0 \\ 1 \\ 0 \\ 0 \\ 0 \\ 0 \\ 0 \\ 0 \\ 0 \end{bmatrix} \oplus \begin{bmatrix} 0 \\ 0 \\ 2 \\ 0 \\ 0 \\ 0 \\ 0 \\ 0 \\ 0 \end{bmatrix} \oplus \begin{bmatrix} 0 \\ 0 \\ 0 \\ 1 \\ 0 \\ 0 \\ 0 \\ 0 \\ 0 \end{bmatrix} \oplus \begin{bmatrix} 0 \\ 0 \\ 0 \\ 0 \\ 2 \\ 0 \\ 0 \\ 0 \\ 0 \end{bmatrix} \oplus \begin{bmatrix} 0 \\ 0 \\ 0 \\ 0 \\ 0 \\ 0 \\ 2 \\ 0 \\ 0 \end{bmatrix} \oplus \begin{bmatrix} 0 \\ 0 \\ 0 \\ 0 \\ 0 \\ 0 \\ 0 \\ 0 \\ 1 \end{bmatrix}$$

$$= 1 \cdot \begin{bmatrix} 0 \\ 1 \\ 0 \\ 0 \\ 0 \\ 0 \\ 0 \\ 0 \\ 0 \end{bmatrix} \oplus 2 \cdot \begin{bmatrix} 0 \\ 0 \\ 1 \\ 0 \\ 0 \\ 0 \\ 0 \\ 0 \\ 0 \end{bmatrix} \oplus 1 \cdot \begin{bmatrix} 0 \\ 0 \\ 0 \\ 1 \\ 0 \\ 0 \\ 0 \\ 0 \\ 0 \end{bmatrix} \oplus 2 \cdot \begin{bmatrix} 0 \\ 0 \\ 0 \\ 0 \\ 2 \\ 0 \\ 0 \\ 0 \\ 0 \end{bmatrix} \oplus 2 \cdot \begin{bmatrix} 0 \\ 0 \\ 0 \\ 0 \\ 0 \\ 0 \\ 2 \\ 0 \\ 0 \end{bmatrix} \oplus 1 \cdot \begin{bmatrix} 0 \\ 0 \\ 0 \\ 0 \\ 0 \\ 0 \\ 0 \\ 0 \\ 1 \end{bmatrix}.$$

This leads to the expression for f as

$$f(x_1, x_2) = 1 \cdot J_0(x_1)J_1(x_2) \oplus 2 \cdot J_0(x_1)J_2(x_2) \oplus 1 \cdot J_1(x_1)J_0(x_2) \oplus 2 \cdot J_1(x_1)J_1(x_2)$$
$$\oplus 2 \cdot J_2(x_1)J_0(x_2) \oplus 1 \cdot J_2(x_1)J_2(x_2).$$

The above example introduces expressions that are an analog of the SOP expressions in binary logic. In matrix notation, the products $\prod J_{q_i}(x_i)$ for $q_i \in \{0, 1, \ldots, p-1\}$, and $i \in \{1, 2, \ldots, n\}$, can be viewed as columns of the $(p^n \times p^n)$ matrix, and by using a signal processing terminology we can say that (2.1) is the decomposition of f in terms of block pulse functions. By selecting functions of different waveforms, we can define many different functional expressions. A reasonable choice would be to have basis functions similar to those already used in the representation of functions in classical mathematical analysis, as for instance by Taylor series or Fourier series. Both these approaches are used in the interpretation of the Reed-Muller expressions for binary functions, and the corresponding extensions to multiple-valued functions will be discussed here under the terms Galois field expressions and Reed-Muller-Fourier expressions, respectively.

2.4   GALOIS FIELD EXPRESSIONS

By adopting the Fourier-oriented approach, the Galois field (GF) expressions over a field of order p can be viewed as the series expressions in terms of a particular set of basis functions consisting of the constant 1 and the set of elementary products of p-valued variables. Thus, basis functions are defined as product of variables in the elements of the power set of the set of n variables and their powers of order up to $p - 1$. It follows that in a GF-expression no products of the same set of variables can appear. This is the same restriction as in the binary case, since the Reed-Muller expressions are the GF-expressions in $GF(2)$.

Definition 10 *(Basis functions for GF-expressions)*
Consider a set of n variables x_i each of them taking values in a finite set $G_i = \{0, 1, \ldots p_i - 1\}$ that is the support set of a finite field $GF(p_i)$. For each variable x_i, consider also its powers x_i^k up to the order $p - 1$, i.e., $k = 1, \ldots, p - 1$, with the exponentiation derived from the multiplication in $GF(p)$. By definition, $x_i^0 = 1$. Consider the power set $P(X)$ of the set X whose elements are variables and their powers.

To each element S_i of $P(X)$, we assign a function ϕ_i defined as the product of elements of S_i with the multiplication in $GF(p_i)$

$$\phi_i(x) = x_1^{i_1} x_2^{i_2} \cdots x_n^{i_n}, \quad x = (x_1, x_2, \ldots, x_n), \quad i = (i_1, i_2, \ldots, i_n). \tag{2.2}$$

It is clear that the function $\phi_0 = 1$. The functions $\phi_i(x)$ define the basis in terms of which the GF-expressions are defined.

Definition 11 *(GF-expressions)*
For a function $f(x) = f(x_1, x_2, \ldots, x_n), x_i \in GF(p)$, the GF-expressions is the polynomial expression of the form

$$f(x) = \sum_{i=1}^{K} g_i \phi_i(x), \tag{2.3}$$

where $K = \prod_{i=1}^{n} p_i$.

In matrix notation, the functions $\phi_i(x)$ are defined as

$$\mathbf{X}_{GF} = \bigotimes_{i=1}^{n} \mathbf{X}_i, \quad \mathbf{X}_i = \begin{bmatrix} x_i^0 & x_i^1 & \cdots & x_i^{p-1} \end{bmatrix}. \tag{2.4}$$

Since being generated as the Kronecker product, the basis functions $\phi_i(x)$ are in the so-called Hadamard ordering.

For each variable x_i, we consider a matrix $\mathbf{B}_i(1)$ whose columns are $x_i^k, k = 0, 1, \ldots p_i - 1$. Thus, $\mathbf{X}_i$ is the symbolic notation for the columns of $\mathbf{B}_i(1)$. Notice that the constant 1 is included as the first column corresponding to x_i^0. We define a matrix

$$\mathbf{G}_i(1) = (\mathbf{B}_i(1))^{-1}, \tag{2.5}$$

where -1 denotes the inverse of the matrix $\mathbf{B}_i(1)$ computed over $GF(p_i)$. For consistency of notation, in the following considerations $\mathbf{B}_i(1)$ will be denoted as $(\mathbf{G}_i(1))^{-1}$ and we will speak about the *direct* and the *inverse* transforms used to determine coefficients in GF-expressions.

Definition 12 *(GF-matrix)*
The Galois field (GF) matrix is defined as

$$\mathbf{G}(n) = \bigotimes_{i=1}^{n} \mathbf{G}(1). \tag{2.6}$$

Definition 13 *(GF-expressions in matrix notation)*
In matrix notation, for a function $f(x) = f(x_1, x_2, \ldots, x_n), x_i \in GF(p)$ specified by the function vector
$\mathbf{F} = [f(0), f(1), \ldots f(K)]^T$, *the GF-expressions is defined as*

$$f(x) = \mathbf{X}(n)\mathbf{S}_f, \tag{2.7}$$

where

$$\mathbf{S}_f(n) = \mathbf{G}(n)\mathbf{F}, \tag{2.8}$$

is the GF-spectrum *for f. Thus,*

$$f(x) = \mathbf{X}(n)\mathbf{G}(n)\mathbf{F}. \tag{2.9}$$

These definitions will be illustrated by the examples for $p = 3$ and $p = 4$.

2.4.1   GALOIS FIELD EXPRESSIONS FOR TERNARY FUNCTIONS

Each n-variable ternary function can be represented as a polynomial of the form

$$f(x_1, \ldots, x_n) = \sum_{i=0}^{3^n-1} g_i \phi_i, \tag{2.10}$$

where $g_i, i \in \{0, 1, 2, 3\}$, and ϕ_i are the product terms defined in the Hadamard order as elements
of the vector $\mathbf{X}_{3GF}(n)$ defined by

$$\mathbf{X}_{3GF}(n) = \bigotimes_{i=1}^{n} \mathbf{X}_{3GF}(1), \quad \mathbf{X}_{3GF}(1) = \begin{bmatrix} x_i^0 & x_i^1 & x_i^2 \end{bmatrix},$$

and addition and multiplication are carried out in $GF(3)$, i.e., modulo 3.

Therefore, when written explicitly, the set of basic functions for $n = 1$ is given by columns of
the matrix

$$\mathbf{X}_{3GF}(1) = \begin{bmatrix} 1 & 0 & 0 \\ 1 & 1 & 1 \\ 1 & 2 & 1 \end{bmatrix}.$$

In matrix notation, for a function f specified by the function vector $\mathbf{F} = [f(0), \ldots, f(3^n - 1)]^T$, the coefficients a_i in the Galois field expression are calculated as

$$\mathbf{S}_{f,3GF} = \mathbf{G}_{3GF}(n)\mathbf{F},$$

where

$$\mathbf{G}_{3GF}(n) = \bigotimes_{i=0}^{n} \mathbf{G}_{3GF}(1), \quad \mathbf{G}_{3GF}(1) = (\mathbf{X}_{3GF}(1))^{-1} = \begin{bmatrix} 1 & 0 & 0 \\ 0 & 2 & 1 \\ 2 & 2 & 2 \end{bmatrix} \quad \text{in } GF(3).$$

With this notation, (2.11) may be written as

$$f(x_1, \ldots, x_n) = \mathbf{X}_{3GF}(n)\mathbf{G}(n)\mathbf{F}.$$

Example 11 *For $n = 2$, the Galois field transform matrix for $GF(3)$ is defined as*

$$\mathbf{G}_{3GF}(2) = \begin{bmatrix} 1 & 0 & 0 \\ 0 & 2 & 1 \\ 2 & 2 & 2 \end{bmatrix} \otimes \begin{bmatrix} 1 & 0 & 0 \\ 0 & 2 & 1 \\ 2 & 2 & 2 \end{bmatrix}$$

$$= \begin{bmatrix} 1 & 0 & 0 & 0 & 0 & 0 & 0 & 0 & 0 \\ 0 & 2 & 1 & 0 & 0 & 0 & 0 & 0 & 0 \\ 2 & 2 & 2 & 0 & 0 & 0 & 0 & 0 & 0 \\ 0 & 0 & 0 & 2 & 0 & 0 & 1 & 0 & 0 \\ 0 & 0 & 0 & 0 & 1 & 2 & 0 & 2 & 1 \\ 0 & 0 & 0 & 1 & 1 & 1 & 2 & 2 & 2 \\ 2 & 0 & 0 & 2 & 0 & 0 & 2 & 0 & 0 \\ 0 & 1 & 2 & 0 & 1 & 2 & 0 & 1 & 2 \\ 1 & 1 & 1 & 1 & 1 & 1 & 1 & 1 & 1 \end{bmatrix}.$$

2.4.2 GALOIS FIELD EXPRESSIONS FOR QUATERNARY FUNCTIONS

Table 2.2 shows addition and multiplication in $GF(4)$ which are operations in terms of which the GF-expressions for quaternary functions are defined.

Table 2.2: Addition and multiplication in $GF(4)$

+	0	1	2	3		$\cdot$	0	1	2	3
0	0	1	2	3		0	0	0	0	0
1	1	0	3	2		1	0	1	2	3
2	2	3	0	1		2	0	2	3	1
3	3	2	1	0		3	0	3	1	2

Each n-variable quaternary function can be represented as a polynomial of the form

$$f(x_1, \ldots, x_n) = \sum_{i=0}^{4^n-1} g_i \phi_i, \tag{2.11}$$

where g_i, $i \in \{0, 1, 2, 3\}$, ϕ_i are the product terms defined in the Hadamard order as elements of the vector $\mathbf{X}_{4GF}(n)$ defined by

$$\mathbf{X}_{4GF}(n) = \bigotimes_{i=1}^{n} \mathbf{X}_{4GF}(1), \quad \mathbf{X}_{4GF}(1) = \begin{bmatrix} x_i^0 & x_i^1 & x_i^2 & x_i^3 \end{bmatrix},$$

and addition and multiplication are carried out in $GF(4)$.

Therefore, when written explicitly, the set of basic functions for $n = 1$ is given by columns of the matrix

$$\mathbf{X}_{4GF}(1) = \begin{bmatrix} 1 & 0 & 0 & 0 \\ 1 & 1 & 1 & 1 \\ 1 & 2 & 3 & 1 \\ 1 & 3 & 2 & 1 \end{bmatrix}.$$

The coefficients in GF-representations (2.11) can be calculated by using:

1. the Newton's divided differences method [145], [247];

2. its modifications [14], [114], [263] intended to improve the efficiency of calculations; or

3. FFT-like algorithms; for example, see [77] derived from the matrix representation of the calculation procedure.

The calculation procedure for coefficients in Galois field representations can be given in the matrix notation as follows.

For a function $f : GF(4)^n \to GF(4)$ given by its function vector $\mathbf{F}(n) = \begin{bmatrix} f(0), \ldots, f(4^n - 1) \end{bmatrix}^T$, the vector of GF-coefficients $\mathbf{S}_{f,4GF} = \begin{bmatrix} g_0, \ldots, g_{4^n-1} \end{bmatrix}^T$ can be determined as

$$\mathbf{S}_{f,4GF} = \mathbf{G}(n)\mathbf{F},$$

where

$$\mathbf{G}_{4GF}(n) = \bigotimes_{i=1}^{n} \mathbf{G}_{4GF}(1), \quad \mathbf{G}_{4GF}(1) = \begin{bmatrix} 1 & 0 & 0 & 0 \\ 0 & 1 & 3 & 2 \\ 0 & 1 & 2 & 3 \\ 1 & 1 & 1 & 1 \end{bmatrix}.$$

With this notation, (2.11) may be written as

$$f(x_1, \ldots, x_n) = \mathbf{X}_{4GF}(n)\mathbf{G}_{4GF}(n)\mathbf{F}.$$

2.5   FIXED-POLARITY GF-EXPRESSIONS

An important property of the GF-transform matrices $\mathbf{G}_i$ is that, unlike the Reed-Muller transform matrix in the binary case, it is not a triangular matrix, which reflects to the optimization of GF-representations.

In the binary case, the logic negation of a variable can be viewed as the permutation of the values a variable can take. This interpretation can be directly generalized to the multiple-valued case by allowing permutation of the values a p-valued variable can take. There are, however, some restrictions that follow from the requirements imposed on the product terms and operations in $GF(p)$.

A consequence of the form of basis functions ϕ_i in terms of which the GF-expressions are defined is that in the optimization by selecting different polarities for variables we are restricted to p out of $p!$ possible permutations of the values a variable can take. These permutations are defined as $\overset{k-}{x} = x \oplus k, k = 1, 2, \ldots, p - 1$. All other permutations do not change the number of non-zero coefficients, which follows from the structure of the GF-transform matrix. This disadvantage in reduction of the number possible different functional expansions is overcome in the Reed-Muller-Fourier transform discussed below.

It is a customary to represent the selection of polarities for variables in GF-expressions in an n-variable p-valued function by the polarity vector $\mathbf{H} = [h_1, \ldots, h_n], h_i \in \{0, 1, 2, \ldots, p - 1\}$, whose i-th coordinate taking the value k determines the polarity $\overset{k-}{x}_i = x_i \oplus k$ of the variable x_i.

Example 12 *Table 2.3 shows the polarity for ternary variables, the corresponding basis functions, and GF-transform matrices. The ternary function $f(x_1, x_2) = x_1 \oplus x_2$ has the function vector $\mathbf{F} = [0, 1, 2, 1, 2, 0, 2, 0, 1]^T$. The GF-coefficients for all FPGF-expressions are shown in Table 2.4. We see that the optimal polarities are $\mathbf{H} = [0, 0]$, $\mathbf{H} = [1, 2]$, $\mathbf{H} = [2, 1]$, and $\mathbf{H} = [2, 2]$, since the corresponding GF-expressions have 2 non-zero coefficients. All other GF-expressions have 3 non-zero coefficients compared with 6 non-zero function values.*

Example 13 *Consider a two-variable ternary function f specified by the function vector $\mathbf{F} = [1, 1, 1, 2, 1, 1, 1, 2, 2]^T$. Table 2.5 shows the GF-spectra and the number of non-zero coefficients for different polarities. The minimum number of GF-coefficients is 4 for the polarities $H = [0, 0]$ and $H = [2, 0]$.*

Notice that it may be opportune to consider the representations with the simplest set of coefficient values, for instance about the representations having as many as possible non-zero coefficients equal to 1. In general, the coefficients take values in the set $\{0, 1, 2, \ldots, p - 1\}$. However, in practical realizations, the coefficients different from 1 represent a multiplication and it can be desirable to reduce their number.

The exact optimization of Fixed-polarity Galois field (FPGFs) expressions consists in the determination of all possible p^n expressions for a given function of n variables and selection of the

Table 2.3: Complements in $GF(3)$ and transform matrices

Variable	Transform	Inverse transform
$x = \begin{bmatrix} 0 \\ 1 \\ 2 \end{bmatrix}$	$(\mathbf{X})(1) = \begin{bmatrix} 1 & 0 & 0 \\ 1 & 1 & 1 \\ 1 & 2 & 1 \end{bmatrix}$	$\mathbf{G}_{3GF}(1) = \begin{bmatrix} 1 & 0 & 0 \\ 0 & 2 & 1 \\ 2 & 2 & 2 \end{bmatrix}$
$\overset{1-}{x} = \begin{bmatrix} 1 \\ 2 \\ 0 \end{bmatrix}$	$\overset{1-}{(\mathbf{X})}(1) = \begin{bmatrix} 1 & 1 & 1 \\ 1 & 2 & 1 \\ 1 & 0 & 0 \end{bmatrix}$	$\overset{1-}{\mathbf{G}}_{3GF}(1) = \begin{bmatrix} 0 & 0 & 1 \\ 2 & 1 & 0 \\ 2 & 2 & 2 \end{bmatrix}$
$\overset{2-}{x} = \begin{bmatrix} 2 \\ 0 \\ 1 \end{bmatrix}$	$\overset{2-}{(\mathbf{X})}(1) = \begin{bmatrix} 1 & 2 & 1 \\ 1 & 0 & 0 \\ 1 & 1 & 1 \end{bmatrix}$	$\overset{2-}{\mathbf{G}}_{3GF}(1) = \begin{bmatrix} 0 & 1 & 0 \\ 1 & 0 & 2 \\ 2 & 2 & 2 \end{bmatrix}$

Table 2.4: GF-coefficients for the function in Example 12

Polarity	GF-spectrum	q
$H = [00]$	$[0, 1, 0, 1, 0, 0, 0, 0, 0]^T$	2
$H = [01]$	$[2, 1, 0, 1, 0, 0, 0, 0, 0]^T$	3
$H = [02]$	$[1, 1, 0, 1, 0, 0, 0, 0, 0]^T$	3
$H = [10]$	$[2, 1, 0, 1, 0, 0, 0, 0, 0]^T$	3
$H = [11]$	$[1, 1, 0, 1, 0, 0, 0, 0, 0]^T$	3
$H = [12]$	$[0, 1, 0, 1, 0, 0, 0, 0, 0]^T$	2
$H = [20]$	$[1, 1, 0, 1, 0, 0, 0, 0, 0]^T$	3
$H = [21]$	$[0, 1, 0, 1, 0, 0, 0, 0, 0]^T$	2
$H = [22]$	$[2, 1, 0, 1, 0, 0, 0, 0, 0]^T$	2

expression with the minimum number of product terms and minimum number of literals in them. The main problem here is the exponential space and time complexity of computations. There are several methods for efficient determination of FPGFs for both binary and multiple-valued functions; for example, see [38], [53], [62], [86], [87], [88], [226], [231], [232], [256], [257].

In particular, methods exploiting the *dual polarity* property have been proven to be reasonably efficient [93], [227]. Efficiency of the determination of such expressions has been further improved by introducing and using the notion of *extended dual polarity property* [91], [92], [93].

In [94], the method of efficient computation of FPGFs is further improved by introducing the notion of the *homogeneous extended dual polarity routes* as a subset of all possible *extended dual polarity routes*. It is shown that in computing FPGF expressions the transition among polarities can be done in the most efficient way along the *homogeneous extended dual polarity routes*. Experimental results support the corresponding method [95].

Table 2.5: GF-coefficients for the function in Example 13

Polarity	GF-spectrum	q
$H = [00]$	$[1, 0, 0, 2, 0, 2, 2, 0, 0]^T$	4
$H = [01]$	$[1, 0, 0, 1, 2, 2, 2, 0, 0]^T$	5
$H = [02]$	$[1, 0, 0, 1, 1, 2, 2, 0, 0]^T$	5
$H = [10]$	$[1, 0, 1, 1, 0, 2, 2, 0, 0]^T$	5
$H = [11]$	$[2, 1, 1, 0, 2, 2, 2, 0, 0]^T$	6
$H = [12]$	$[2, 2, 1, 0, 2, 2, 2, 0, 0]^T$	6
$H = [20]$	$[2, 0, 2, 0, 0, 2, 2, 0, 0]^T$	4
$H = [21]$	$[1, 2, 2, 2, 2, 2, 2, 0, 0]^T$	7
$H = [22]$	$[1, 1, 2, 2, 1, 2, 2, 0, 0]^T$	7

2.5.1 REED-MULLER-FOURIER TRANSFORM

GF-expressions share some of the properties of Fourier series for real-valued functions of real-valued variables, however, there is a considerable discrepancy among other properties. At the same time, the transform matrix used in GF-expressions does not have a triangular form as it is the case in the Reed-Muller expressions of binary functions. These two reasons motivated to change the underlying algebraic structure and introduce the Gibbs algebra allowing to consider the Reed-Muller expressions as the Instant Fourier transform [68]. A generalization to multiple-valued functions was introduced in [184], leading to the Reed-Muller-Fourier (RMF) expressions [205], [209]. In this section, we present basic definitions and a brief theory of RMF-expressions for $p = 3$ and $p = 4$. Note that for $p = 2$, these expressions reduce to the Reed-Muller expressions for binary logic functions.

Denote by G a group of n-ary p-valued sequences $x = (x_1, \ldots, x_n)$ with group operation defined as componentwise addition modulo p. Thus, for all $x = (x_1, \ldots, x_n)$, $y = (y_1, \ldots, y_n) \in G$,

$$\begin{aligned} x \oplus y &= (x_1, \ldots x_n) \oplus (y_1, \ldots, y_n) \\ &= ((x_1 \oplus y_1), \ldots, (x_n \oplus y_n)) \bmod p. \end{aligned}$$

Denote by Z_q the set of first q non-negative integers. For each $x \in G$, the p-adic contraction is defined as a mapping $\sigma : G \to Z_q$ given by

$$\sigma(x) = \sum_{i=1}^{n} x_i p^{n-i}.$$

We denote by $P(G)$ the set of all functions $f : G \to Z_q$. In $P(G)$, we define the addition as modulo p addition,

$$(f \oplus g)(x) = f(x) \oplus g(x), \forall x \in G,$$

and multiplication as a convolutionwise (Gibbs) multiplication [68]

$$(fg)(0) = 0$$
$$(fg)(x) = \sum_{s=0}^{\sigma(x)-1-s} f(\sigma(x) - 1 - s)g(s), \forall x \in G, x \neq 0.$$

Denote by W a particular function in $P(G)$ such that

$$W(x) = p - 1, \quad \forall x \in G,$$

and by S the set of first q positive integer powers of W, i.e., $S = \{W^1, \ldots, W^q\}$. The set S is a basis in $P(G)$ with respect to which the Reed-Muller-Fourier (RMF) transform is defined as [184]

$$f = \sum_{i=0}^{q-1} c_i W^{i+1} \bmod p,$$

where $c_i \in Z_p$.

As is shown in [184], if a p-valued variable x_i is considered as a particular function in $P(G)$, $f(x_1, \ldots, x_n) = x_i$, then the RMF-transform matrix can be expressed in terms of the variable x_n as

$$f(x_1, \ldots, x_n) = \sum_{i=0}^{q-1} c_i \phi_{i+1}(x_n) \bmod p, c_i \in \{0, \ldots, p - 1\},$$

where

$$\phi_i(x_n) = \begin{cases} (p - 1) \cdot x_n^{\lceil i/2 \rceil}, & i\text{-odd}, \\ x_n^{i/2}, & i\text{-even}, \end{cases}$$

where $\lceil a \rceil$ is the integer part of a, and $x_n^r = x_n \ldots x_n$ r times with the multiplication defined as convolutionwise (Gibbs) multiplication in $P(G)$.

This definition corresponds to the interpretation of the RMF-expressions as an analog of the Fourier series expressions with functions $\phi_i(x_n)$ viewed as counterparts of the exponential functions. The following alternative definition allows to consider the RMF-expressions as polynomial expressions representing a generalization of the Reed-Muller expressions for binary logic functions to multiple-valued functions.

Definition 14 *(Reed-Muller-Fourier expressions)*
Any p-valued n-variable function $f(x_1, \ldots, x_n)$ can be expanded in powers of variables $x_i, i = 1, \ldots, n$ as

$$f(x_1, \ldots, x_n) = (-1)^n \sum_{a \in V^n} q(a) x_1^{*a_1} \cdots x_n^{*a_n},$$

where V^n is the set of all p-valued n-tuples, $q(a) \in \{0, 1, 2, \ldots p - 1\}$, and the exponentiation is defined as $x^{*0} = -1$ modulo p, and for $i > 0$, x^{*i} is determined in terms of the convolutionwise (Gibbs) multiplication defined above.

In what follows, we provide case examples for $p = 3$ and $p = 4$.

2.5.2   RMF EXPRESSIONS FOR $p = 3$.

Consider the ring of integers modulo 3 defined in terms of addition and multiplication modulo 3 given by Table 2.6. In order to generate the product terms of three-valued variables corresponding to that appearing in the RM-expressions for switching functions and GF-expressions for multiple-valued functions, we define in Table 2.7 the 3AND multiplication and 3EXP exponentiation, denoted by $\odot$ and $*$, respectively. Note that 3AND table is actually the multiplication modulo 3 table multiplied by 2. Notice that $2 = -1$ modulo 3.

We generate a set of 3^n product terms given in the matrix notation by

$$\mathbf{X}_{3RMF}(n) = \bigotimes_{i=1}^{n} \begin{bmatrix} x_i^{*0} & x_i^{*1} & x_i^{*2} \end{bmatrix}$$

$$= \bigotimes_{i=1}^{n} \begin{bmatrix} 2 & x_i & x_i^{*2} \end{bmatrix},$$

with 3AND and 3EXP applied to the three-valued variables. In matrix notation, the basis functions are expressed as column of the matrix

$$\mathbf{X}_{3RMF}(1) = \begin{bmatrix} 2 & 0 & 0 \\ 2 & 1 & 0 \\ 2 & 2 & 2 \end{bmatrix}.$$

Table 2.6: Addition and multiplication modulo 3

$\oplus$	0	1	2		$\cdot$	0	1	2
0	0	1	2		0	0	0	0
1	1	2	0		1	0	1	2
2	2	0	1		2	0	2	1

Table 2.7: 3EXP and 3AND

$*$	0	1	2		$\odot$	0	1	2
0	2	0	0		0	0	0	0
1	2	1	0		1	0	2	1
2	2	2	2		2	0	1	2

Definition 15 *Each n-variable 3-valued logic function given by the truth-vector $\mathbf{F}(n) = [f(0), \ldots, f(3^n - 1)]^T$ can be represented as a Reed-Muller-Fourier (RMF) polynomial given by*

$$f(x_1, \ldots x_n) = (-1)^n \mathbf{X}_{3RMF}(n) \mathbf{S}_{f,3RMF}(n),$$

with calculations modulo 3, where the vector of RMF-coefficients $\mathbf{S}_{f,3RMF}(n) = [a_0, \ldots, a_{3^n-1}]^T$ *is determined by the matrix relation*

$$\mathbf{S}_{f,3RMF}(n) = \mathbf{R}_{3RMF}(n)\mathbf{F}(n),$$
$$\mathbf{R}_{3RMF}(n) = \bigotimes_{i=1}^{n} \mathbf{R}_{3RMF}(1),$$

where

$$\mathbf{R}_{3RMF}(1) = \mathbf{X}_{3RMF}^{-1}(1) = 2 \begin{bmatrix} 1 & 0 & 0 \\ 1 & 2 & 0 \\ 1 & 1 & 1 \end{bmatrix}.$$

Note that $\mathbf{X}_{3RMF}^{-1}(1)$ is its own inverse.

Example 14 *For $p = 3$ and $n = 2$, the RMF-transform matrix is*

$$\mathbf{R}_{3RMF}(2) = 2 \begin{bmatrix} 1 & 0 & 0 \\ 1 & 2 & 0 \\ 1 & 1 & 1 \end{bmatrix} \otimes 2 \begin{bmatrix} 1 & 0 & 0 \\ 1 & 2 & 0 \\ 1 & 1 & 1 \end{bmatrix}$$

$$= \begin{bmatrix} 1 & 0 & 0 & 0 & 0 & 0 & 0 & 0 & 0 \\ 1 & 2 & 0 & 0 & 0 & 0 & 0 & 0 & 0 \\ 1 & 1 & 1 & 0 & 0 & 0 & 0 & 0 & 0 \\ 1 & 0 & 0 & 2 & 0 & 0 & 0 & 0 & 0 \\ 1 & 2 & 0 & 2 & 1 & 0 & 0 & 0 & 0 \\ 1 & 1 & 1 & 2 & 2 & 2 & 0 & 0 & 0 \\ 1 & 0 & 0 & 1 & 0 & 0 & 1 & 0 & 0 \\ 1 & 2 & 0 & 1 & 2 & 0 & 1 & 2 & 0 \\ 1 & 1 & 1 & 1 & 1 & 1 & 1 & 1 & 1 \end{bmatrix} = \mathbf{X}_{3RMF}(2).$$

The basis functions used to define the RMF-expressions for $p = 3$ are defined as

$$2\mathbf{X}_{3RMF}(2) = [2, x_2, x_2^{*2}, x_1, x_1 \odot x_2 \\ x_1 \odot x_2^{*2}, x_1^{*2}, x_1^{*2} \odot x_2, x_1^{*2} \odot x_2^{*2}].$$

Example 15 *For the ternary function $f(x_1, x_2) = x_1 \oplus x_2$, specified by the function vector $\mathbf{F} = [0, 1, 2, 1, 2, 0, 2, 0, 1]^T$, the RMF-spectrum is* $\mathbf{S}_{f,3RMF} = [0, 2, 0, 2, 0, 0, 0, 0, 0]^T$.

Due to its Kronecker product structure, the RMF-matrix $\mathbf{R}_{3,RMF}(n)$ can be defined recursively as

$$\mathbf{R}_{3,RMF}(n) = 2 \begin{bmatrix} \mathbf{R}_{3,RMF}(n-1) & \mathbf{0}(n-1) & \mathbf{0}(n-1) \\ \mathbf{R}_{3,RMF}(n-1) & 2\mathbf{R}_{3,RMF}(n-1) & \mathbf{0}(n-1) \\ \mathbf{R}_{3,RMF}(n-1) & \mathbf{R}_{3,RMF}(n-1) & \mathbf{R}_{3,RMF}(n-1) \end{bmatrix}.$$

2.5.3 RMF EXPRESSIONS FOR $p = 4$.

Consider the ring of integers modulo 4 defined in terms of addition and multiplication modulo 4 given by Table 2.8. In order to generate the product terms of four-valued switching variables corresponding to that appearing in the RM-expressions for switching functions and GF-expressions for multiple-valued functions, we define in Table 2.9 the 4AND multiplication and 4EXP exponentiation, denoted by $\odot$ and $*$, respectively. Note that 4AND table is actually the multiplication modulo 4 table multiplied by 3. Notice that $3 = -1$ modulo 4.

Table 2.8: Addition and multiplication modulo 4

$\oplus$	0	1	2	3	$\cdot$	0	1	2	3
0	0	1	2	3	0	0	0	0	0
1	1	2	3	0	1	0	1	2	3
2	2	3	0	1	2	0	2	0	2
3	3	0	1	2	3	0	3	2	1

Table 2.9: 4EXP and 4AND

$*$	0	1	2	3	$\odot$	0	1	2	3
0	3	0	0	0	0	0	0	0	0
1	3	1	0	0	1	0	3	2	1
2	3	2	3	0	2	0	2	0	2
3	3	3	1	1	3	0	1	2	3

We generate a set of 4^n product terms given in the matrix notation by

$$\mathbf{X}_{4RMF}(n) = \bigotimes_{i=1}^{n} \left[\; x_i^{*0} \quad x_i^{*1} \quad x_i^{*2} \quad x_i^{*3}, \; \right]$$

$$= \bigotimes_{i=1}^{n} \left[\; 3 \quad x_i \quad x_i^{*2} \quad x_i^{*3}, \; \right],$$

with 4AND and 4EXP applied to the four-valued variables. In matrix notation, the basis functions represented as columns of a matrix are

$$\mathbf{X}_{4RMF}(1) = \begin{bmatrix} 3 & 0 & 0 & 0 \\ 3 & 1 & 0 & 0 \\ 3 & 2 & 3 & 0 \\ 3 & 3 & 1 & 1 \end{bmatrix}.$$

Definition 16 *Each n-variable 4-valued logic function given by the truth-vector* $\mathbf{F}(n) = [f(0), \ldots, f(4^n - 1)]^T$ *can be represented as a Reed–Muller–Fourier (RMF) polynomial given by*

$$f(x_1, \ldots x_n) = (-1)^n \mathbf{X}_{4RMF}(n) \mathbf{S}_f(n),$$

with calculations modulo 4, where the vector of RMF-coefficients $\mathbf{R}_f(n) = [a_0, \ldots, a_{4^n-1}]^T$ *is determined by the matrix relation*

$$\mathbf{S}_{f,4RMF}(n) = \mathbf{R}_{4RMF}(n)\mathbf{F}(n),$$
$$\mathbf{R}_{4RMF}(n) = \bigotimes_{i=1}^{n} \mathbf{R}_{4RMF}(1),$$

where

$$\mathbf{R}_{4RMF}(1) = 3 \begin{bmatrix} 1 & 0 & 0 & 0 \\ 1 & 3 & 0 & 0 \\ 1 & 2 & 1 & 0 \\ 1 & 1 & 3 & 3 \end{bmatrix} = (\mathbf{X}_{4RMF}(1))^{-1}.$$

Notice that $\mathbf{X}_{4RMF}(1)$ *is its own inverse over* $GF(4)$.

Example 16 *For n = 2 the vector* $\mathbf{X}_{4RMF}(2)$ *of product terms appearing in an RMF-representation is given by*

$$
\begin{aligned}
\mathbf{X}_{4RMF}(2) &= \begin{bmatrix} 1 & x_1 & x_1^{*2} & x_1^{*3} \end{bmatrix} \otimes \begin{bmatrix} 1 & x_2 & x_2^{*2} & x_2^{*3} \end{bmatrix} \\
&= [1 \quad x_2 \quad x_2^{*2} \quad x_2^{*3} \quad x_1 \quad x_1 \odot x_2 \quad x_1 \odot x_2^{*2} \quad x_1 \odot x_2^{*3} \\
&\quad\; x_1^{*2} \quad x_1^{*2} \odot x_2 \quad x_1^{*2} \odot x_2^{*2} \quad x_1^{*2} \odot x_2^{*3} \quad x_1^{*3} \quad x_1^{*3} \odot x_2 \quad x_1^{*3} \odot x_2^{*2} \quad x_1^{*3} \odot x_2^{*3}].
\end{aligned}
$$

The product terms in the Example 16 are used to define GF-expressions for quaternary functions for $n = 2$. The generalization to functions of an arbitrary number of variables is straightforward.

Example 17 *The RMF-expression for a two-variable quaternary function* f *specified by the function vector* $\mathbf{F} = [0, 0, 0, 0, 0, 1, 3, 2, 0, 3, 2, 1, 0, 2, 1, 3]^T$ *is given by*

$$
\begin{aligned}
f(x_1, x_2) &= (x_1 \odot x_2) \oplus 3(x_1 \odot x_2^{*2}) \oplus 3(x_1^{*2} \odot x_2) \\
&\quad \oplus 2(x_1^{*2} \odot x_2^{*2}) \oplus 2(x_1^{*3} \odot x_2^{*3}),
\end{aligned}
$$

since $\mathbf{S}_{f,4RMF} = [0, 0, 0, 0, 0, 1, 3, 0, 0, 3, 2, 0, 0, 0, 0, 2]^T$.

Example 18 *The Reed–Muller–Fourier spectrum of the two-variable function* f *given by the function vector* $\mathbf{F} = [2, 0, 0, 1, 3, 3, 3, 2, 3, 3, 3, 0, 0, 1, 1, 2]^T$ *is* $\mathbf{S}_{f,4RMF} = [2, 2, 2, 1, 3, 2, 2, 0, 3, 2, 2, 2, 2, 3, 3, 1]^T$.

Example 19 *For a two-variable quaternary function* $f(x_1, x_2)$, $x_1, x_2 \in \{0, 1, 2, 3\}$, *specified by the function vector* $\mathbf{F} = [0, 0, 0, 0, 0, 1, 3, 2, 0, 3, 2, 1, 0, 2, 1, 3]^T$, *the RMF-spectrum is* $\mathbf{S}_{f,4RMF} = [0, 0, 0, 0, 0, 1, 3, 0, 0, 3, 2, 0, 0, 0, 0, 2]^T$.

As in the case of Reed-Muller expressions of switching functions and Galois field expressions of multiple-valued functions, different polarity Reed-Muller-Fourier representations of multiple-valued functions can be distinguished. However, unlike the GF-expressions, the structure of the RMF-transform matrix $\mathbf{R}(n)$ permits the consideration of an extended set of complements of multiple-valued variables. If the complement $\bar{x}$ of the binary variable $x \in \{0, 1\}$ is understood as a reordering of elements of the domain $GF(2)$, the concept of the complement can be extended to multiple-valued functions as a possible reordering of elements of $GF(p)$. In that way, for a p-valued variable x the total of $p!$ different complements $\overset{i-}{x}$, $i = 1, \ldots, (p!) - 1$ can be defined permitting the total of $(p!)^n$ different RMF-representations of an n-variable function.

In the considered case $p = 4$, the possible orderings of elements of $GF(4)$ $\overset{i-}{x}$, $i = 1, \ldots, 23$, are given in Table 2.10.

	Table 2.10:		Complements	of	variables	for	RMF-expressions				
x	$\overset{1-}{x}$	$\overset{2-}{x}$	$\overset{3-}{x}$	$\overset{4-}{x}$	$\overset{5-}{x}$	$\overset{6-}{x}$	$\overset{7-}{x}$	$\overset{8-}{x}$	$\overset{9-}{x}$	$\overset{10-}{x}$	$\overset{11-}{x}$
0	1	2	3	0	0	0	0	0	1	1	1
1	0	3	2	1	2	2	3	3	0	2	2
2	3	0	1	3	1	3	1	2	2	0	3
3	2	1	0	2	3	1	2	1	3	3	0
$\overset{12-}{x}$	$\overset{13-}{x}$	$\overset{14-}{x}$	$\overset{15-}{x}$	$\overset{16-}{x}$	$\overset{17-}{x}$	$\overset{18-}{x}$	$\overset{19-}{x}$	$\overset{20-}{x}$	$\overset{21-}{x}$	$\overset{22-}{x}$	$\overset{23-}{x}$
1	1	2	2	2	2	2	3	3	3	3	3
3	3	0	0	1	1	3	0	0	1	1	2
0	2	1	3	0	3	1	1	2	0	2	0
2	0	3	1	3	0	0	2	1	2	0	1

The first four of them are equal to complements of x_i defined as $x_i + i$, where addition is in $GF(4)$. The other are extended complements that can be used in RMF-expressions. Their use in GF-expressions does not reduce the number of non-zero coefficients.

The justification for considering these extended complements is the existence of some functions for which the use of these complements provides simpler representations than both GF and RMF-expressions with complements allowed in GF-expressions.

By using these different complements of variables the optimal Reed-Muller-Fourier expressions with respect to the number of coefficients can be determined as in the case of RM-expressions and Galois field expressions.

Example 20 *For the function* f *specified by the function vector* $\mathbf{F} = [2, 0, 0, 1, 3, 3, 3, 2, 3, 3, 3, 0, 0, 1, 1, 2]^T$, *the zero-polarity spectrum is* $\mathbf{S}_{f,4RMF} = $

$[2, 2, 2, 1, 3, 2, 2, 0, 3, 2, 2, 2, 2, 3, 3, 1]^T$. *The spectrum for the polarity* $H = [0, 14]$ *is* $\mathbf{S}_{f,4RMF} = [0, 0, 2, 1, 1, 0, 2, 0, 1, 0, 2, 2, 3, 0, 3, 3]^T$. *To compute the spectrum, the columns* $\mathbf{r}_0, \mathbf{r}_1, \mathbf{r}_2, \mathbf{r}_3$, *of the basic RMF-matrix* $\mathbf{R}_{4RMF}(1)$ *for the variable* x_2 *are reordered as* $\mathbf{r}_0, \mathbf{r}_1, \mathbf{r}_2, \mathbf{r}_3 \rightarrow \mathbf{r}_2, \mathbf{r}_0, \mathbf{r}_1, \mathbf{r}_3$.

Example 21 *The optimal polarity RMF-expression of the function f specified by the function vector* $\mathbf{F} = [2, 0, 0, 1, 3, 3, 3, 2, 3, 3, 3, 0, 0, 1, 1, 2]^T$ *in Example 18 is obtained for* $H = [0, 19]$ *and is given by* $\mathbf{S}_{f,4RMF} = [0, 0, 1, 1, 1, 0, 2, 0, 0, 0, 1, 1, 0, 0, 1, 0]^T$. *It requires* 7 *instead* 15 *non-zero coefficients in zero-polarity RMF-expressions. In this case, the columns of the basic RMF-matrix* $\mathbf{R}_{4RMF}(1)$ *for the variable* x_2 *are reordered as* $\mathbf{r}_0, \mathbf{r}_1, \mathbf{r}_2, \mathbf{r}_3 \rightarrow \mathbf{r}_3, \mathbf{r}_0, \mathbf{r}_1, \mathbf{r}_2$.

2.6 EFFICIENCY OF REPRESENTATIONS

Efficiency of an analytical representation of switching functions can be considered from two different aspects:

1. Regarding the complexity of calculation of the considered representation. In the case of polynomial representations that means the calculation of the coefficients of the representation.

2. Relative to the complexity of realization of the given function starting from this particular representation. In the case of polynomial representations, that means the number of non-zero coefficients, since each product term in the representation requires a circuit in the network.

These two aspects can be denoted as *computational efficiency* and *realization efficiency* of the representation, respectively.

2.6.1 COMPUTATIONAL EFFICIENCY

As is noted in [264], for any function $f : GF(k) \rightarrow GF(k)$ where $k = p^n$, any power of a prime p, the coefficients a_i, $i = 0, \ldots, k - 1$ of its Galois field representation can be calculated through the use of the Newton's divided difference method [145], [247]. The method is general in the sense that it can be applied to any finite field and requires $\frac{7}{2}k^2 - \frac{11}{2}k + 2$ finite field operations. The use of Menger's theorem [114] permits a method requiring summations over all the elements for every coefficient of the polynomial requiring $3k^2 - 8k + 6$ finite field operations.

A computationally more efficient method is proposed in [264] for the fields of size 2, 3, and 4, requiring $3k^2 - 9k + 8$ finite field operations. The number of operations for the determination of Galois field representations for four-valued functions by using these methods is given in Table 2.11 together with the number of calculations by using the fast algorithms [77] and compared to the number of calculations for Reed-Muller-Fourier expressions. Compared to the Galois field representations of four-valued functions, regarding the number of multiplications and additions, the Reed-Muller-Fourier expressions offer some advantage [201], [204].

Table 2.11: Number of arithmetic operations to compute polynomial representations, $n = 1$

k	NDD	M	ZV	FGF	RMF
2	5	2	2	1	1
3	17	9	8	7	4
4	36	22	20	11	11
5	62	41	-	26	23
7	135	97	-	63	46

NDD - Newton's divided difference method [247],
M - Method based upon the use of Menger's theorem [114],
FGF - Fixed-polarity GF-expressions
ZV - Method proposed by Žilić and Vranešić [264],
RMF - Reed-Muller-Fourier coefficients
k - order of the field

Table 2.12: Calculations for functions of n variables

n	ZV	GF	RMF
1	20	11	11
2	160	88	96
3	960	528	480
4	5120	2816	2816
5	25600	14080	12800

It should be noted that in the Reed-Muller-Fourier expressions the ring operations, i.e., the addition and multiplication modulo 4 are used, while in other algorithms the field operations of $GF(4)$ are assumed.

Note that we need 10 operations to calculate the RMF-expressions followed by the multiplication of the resulting vector with the scaling factors 3 that can be performed as a vector operation within the time equal to that of one multiplication. Therefore, the savings for one-variable functions equal to the difference in performing the Galois field and modulo 4 operations. However, the computational efficiency of RMF-expressions becomes more obvious for a greater number of variables. In Table 2.12 we compare the calculation efficiency of GF-expressions calculated by the ZV-method and by fast algorithms for the calculation of RMF-expressions for different number of variables. Recall that the generalization of the ZV-method to n variable functions reduces to $n4^{n-1}$ times the implementation of the procedure for single variable functions.

Table 2.13: The distribution of single variable quaternary functions realizable within a given number i of coefficients in zero-polarity GF and RMF expressions

i	GF	RMF
0	1	1
1	12	12
2	54	54
3	108	108
4	81	81

Table 2.14: The distribution of single variable quaternary functions realizable within a given number i of coefficients in minimal-polarity GF and RMF expressions

i	GF	RMF
0	1	1
1	39	31
2	90	114
3	126	106
4	0	4

2.6.2   REALIZATION EFFICIENCY

It may be said that an average realization efficiency of RMF-expressions expressed through the amount of functions realizable with a given number of coefficients is considerably greater than for GF-expressions. That is obvious even from the simplest example of single variable functions. The distribution of single variable quaternary functions realizable within a given number i, $i \in \{0, 1, 2, 3, 4\}$ of coefficients is shown in Table 2.13 for zero-polarity and in Table 2.14 for minimal-polarity and compared to the corresponding GF-expressions. It may be seen that the average complexity for zero-polarity is the same for both GF and RMF-expressions. However, 44.53% of the total of 256 functions can be realized with minimal-polarity RMF-expressions with two coefficients, compared to 35.16% of the functions in GF-expressions. In GF-expressions 49.22% of the functions require three coefficients compared to 41.41% of the functions in RMF-expressions.

Regarding the realization efficiency of a particular given function, it is hard to give some concrete conclusions in general form. There are examples where the efficiency of Galois field representations is greater than the Reed-Muller-Fourier representations and vice versa, regarding the number of non-zero coefficients. The following are two illustrative examples.

Example 22 *The Galois field representation of the two–variable function f given by the function vector* $\mathbf{F} = [0, 0, 0, 0, 0, 1, 3, 2, 0, 3, 2, 1, 0, 2, 1, 3]^T$ *is given by*

$$f(x_1, x_2) = x_1^2 x_2^2.$$

However, the Reed-Muller-Fourier expression requires six non–zero coefficients,

$$f(x_1, x_2) \;=\; 3(x_1 \odot x_2) \oplus (x_1 \odot x_2^{*2}) \oplus (x_1^{*2} \odot x_2) \oplus 2(x_1^{*2} \odot x_2^{*2}) \oplus 2(x_1^{*3} \odot x_2^{*3}).$$

Example 23 *The zero-polarity Galois field representation of the simple function for $n = 1$ given by the function vector* $\mathbf{F} = [0, 0, 1, 0]^T$ *is*

$$f(x) = 3x \oplus 2x^2 \oplus x^3,$$

while the zero-polarity Reed–Muller–Fourier expression is given by

$$f(x) = 3x^{*2} \oplus x^{*3}.$$

The optimal polarity Galois field representation is obtained for $H = (2)$ and is given by

$$f(x) = 1 \oplus x^{2 - {*3}},$$

while the optimal polarity Reed–Muller–Fourier expression for $H = (1)$ requires one coefficient and is given by

$$f(x) = x^{1 - {*3}}.$$

The number of products needed to represent various particular two-variable functions by GF and RMF-expressions are compared in Table 2.15. Note that in this table the fixed-polarity expressions do not include the zero-polarity expressions, thus, there are examples where the zero-polarity expressions require the minimum number of coefficients.

Table 2.15: Number of products to represent various two-variable functions

Function	zero-polarity		min-polarity	
	GF	RMF	GF	RMF
$x_1 \oplus x_2$	14	7	14	7
$\overline{x}_1 \overline{x}_2$ GF	14	3	14	3
$\overline{x}_1 \overline{x}_2$ mod 4	12	6	13	6
$x_1 + x_2$	12	7	12	7
$max\{x_1, x_2\}$	13	7	13	7
$min\{x_1, x_2\}$	13	6	13	6
$max\{\overline{x}_1, x_2\}$	13	7	14	10
$min\{\overline{x}_1, x_2\}$	14	8	14	9
$\overline{x}$ defined by the rule $x \oplus \overline{x} = 0$.				

The following experiments compare the efficiency of GF and RMF-expressions for two- and three-variable quaternary functions.

First, we consider two-variable quaternary functions. We generated 20 sets of 1000 two-variable quaternary functions each by using the pseudo-random generator in the standard C compiler. Then we calculate the average number of non-zero coefficients in optimal-polarity GF and RFM-representations for each of these sets (Table 2.16). The average number of non-zero coefficients in GF-representations range from 12.450 up to 12.574. For the RMF-expressions the average number of terms ranges from 7.261–7.446. Therefore, considered over these 20 sets of arbitrary generated functions the average number of non-zero coefficients is 12.547 for GF and 7.367 for RMF-expressions. Thus, in this experiment, the savings in RMF-expressions are at about 41%.

Next, we consider the three-variable quaternary functions and repeated the same experiment (Table 2.17). The average number of non-zero coefficients in GF-representations range from 44.013 up to 44.198. For RMF the average number of terms range from 35.009–35.072. Therefore, the average number of non-zero coefficients is 44.129 for GF and 35.032 for RMF-expressions. Thus, in this experiment, the savings in RMF-expressions are at about 20%.

Table 2.16: Average number of products $p = 4, n = 2$		
	GF	RMF
1.	12.552	7.444
2.	12.554	7.261
3.	12.561	7.361
4.	12.540	7.389
5.	12.571	7.438
6.	12.574	7.369
7.	12.541	7.345
8.	12.504	7.363
9.	12.563	7.446
10.	12.535	7.343
11.	12.547	7.369
12.	12.537	7.359
13.	12.450	7.357
14.	12.574	7.293
15.	12.571	7.341
16.	12.543	7.335
17.	12.531	7.407
18.	12.546	7.351
19.	12.574	7.348
20.	12.565	7.419
av.	12.547	7.367

Table 2.17: Average number of products $p = 4, n = 3$		
	GF	RMF
1.	44.172	35.018
2.	44.063	35.035
3.	44.101	35.057
4.	44.198	35.021
5.	44.158	35.032
6.	44.061	35.027
7.	44.113	35.072
8.	44.186	35.009
9.	44.175	35.020
10.	44.013	35.020
11.	44.156	35.067
12.	44.148	35.031
13.	44.178	35.024
14.	44.024	35.006
15.	44.159	35.056
16.	44.137	35.032
17.	44.192	35.011
18.	44.035	35.019
19.	44.132	35.050
20.	44.192	35.026
av.	44.129	35.032

Table 2.18 compares the number of non-zero coefficients in some benchmark functions. Each output of a multi-output function is considered as a particular quaternary multiple-valued function and represented separately. For this set of benchmark functions and the way of representation described above, GF and RMF-expressions require on the average 3605.57 and 2589.57 products, respectively. Thus, in this example, RMF-expressions require 28.18% fewer products.

Table 2.18: Number of products in GF and RMF-representations		
	GF	RMF
alu4-3	9696	6301
alu4-4	8609	6393
alu4-5	8515	6070
alu4-6	3117	1553
alu4-7	9308	6668
alu4-8	9266	6596
rd84-1	36	130
rd84-2	8	32
rd84-3	81	1
rd84-4	150	131
sao2-1	350	340
sao2-2	338	646
sao2-3	494	699
sao2-4	510	694
av.	3605.57	2589.57

Table 2.19 compares the number of non-zero coefficients in GF and RMF-expressions of n-bit adders for $n = 4, 5, 6, 7$. Again, outputs are represented separately.

On average, for this set of adders and the way of representation described above, GF and RMF-expressions require 329.92 and 82.62 products, respectively. Thus, in this example, RMF-representations require 75% fewer products.

Table 2.20 compares the number of non-zero coefficients in GF and RMF-expressions of n-bit multipliers for $n = 4, 5, 6, 7$.

On average, for this set of multipliers, GF-expressions and RMF-expressions require 1799.61 and 1048.86 products, respectively. Thus, in this example, RMF-expressions require 41.72% fewer products.

A detailed analysis and comparison of efficiency of various expressions including GF and RMF expressions is presented in [2], [3], [4], with a detailed discussion of features of these expressions in [1].

2.7   ARITHMETIC EXPRESSIONS FOR MULTIPLE-VALUED FUNCTIONS

When representing multi-output binary logic functions, a separate Reed-Muller expression is required for each output. Alternatively, k-outputs of a multiple-output function can be viewed as binary

Table 2.19: Number of products in GF and RMF-expressions of n-bit adders

	GF	RMF
add4-1	54	26
add4-2	32	39
add4-3	10	15
add4-4	8	10
add4-5	4	5
add5-1	391	110
add5-2	196	81
add5-3	62	45
add5-4	22	21
add5-5	8	10
add5-6	4	5
add6-1	438	138
add6-2	224	167
add6-3	58	61
add6-4	32	39
add6-5	10	15
add6-6	8	10
add6-7	4	5
add7-1	4087	574
add7-2	2014	385
add7-3	620	209
add7-4	196	97
add7-5	62	45
add7-6	22	21
add7-7	8	10
add7-8	4	5
av.	329.92	82.62

Table 2.20: Number of non-zero coefficients in GF and RMF-representations of multipliers

	GF	RMF
mul6-1	1731	410
mul6-2	1652	992
mul6-3	2619	1305
mul6-4	2362	1538
mul6-5	2314	1392
mul6-6	1177	924
mul6-7	616	374
mul6-8	102	107
mul6-9	65	37
mul6-10	15	15
mul6-11	2	5
mul6-12	4	3
mul7-1	6083	2047
mul7-2	7970	3920
mul7-3	9424	5563
mul7-4	10530	6817
mul7-5	10226	7012
mul7-6	9725	6110
mul7-7	5876	3870
mul7-8	2505	1327
mul7-9	620	427
mul7-10	197	130
mul7-11	61	42
mul7-12	18	15
mul7-13	8	7
mul7-14	4	3
av.	1799.61	1048.86

encoding of integers which can be represented by k bits. In this way, a multiple-output function is identified with an integer function which can be represented by the arithmetic expressions defined as integer counterpart of the Reed-Muller expressions [32], [49], [78], [111], [112], [144], [248]. This means we keep the same set of basis functions as determined by the primary products of binary variables, or columns of the Reed-Muller matrix $\mathbf{R}(n)$, however, with function values interpreted as integers 0 and 1 instead of logic values. This matrix we denote by $\mathbf{A}^{-1}(n)$. We take the inverse of it over the field of rational numbers Q, as the arithmetic transform matrix $\mathbf{A}(n)$ which is used to define the coefficients in the arithmetic expressions. Since coefficients in arithmetic expressions are integers, which means computer words are required to represent them, these expressions belong to the broad class of various word-level functional expressions for binary-valued functions.

Since the arithmetic expressions are defined with respect to the same set of basis functions as the Reed-Muller expressions, the optimization of arithmetic expressions is performed by selecting polarities of variables in the same way as in the Reed-Muller expressions. In this way, the Fixed-polarity arithmetic expressions are defined [111], [112]. In this context, see also [168], [192].

The generalization to multiple-valued functions is straightforward in the sense that we can consider p-valued k-tuples as multiple-valued representations of integers. In this case, we have more opportunities than in the binary case, since we can discuss generalizations with respect to either GF-expressions or RMF-expressions.

2.7.1   ARITHMETIC EXPRESSIONS FOR MULTIPLE-VALUED FUNCTIONS DERIVED FROM THE GF-EXPRESSIONS

Extensions of GF-expressions for multiple-valued functions to the corresponding word-level expressions can be done in two different ways. In the first approach, the definition of the basis functions is retained, i.e., basis functions are defined as products of integer powers of variables x_i^k, $k \in \{0, 1, \ldots, p-1\}$, where exponentiation is derived from multiplication in the field of rational numbers Q. In the second approach, the form of basis functions is preserved, i.e., we use the same basis functions as in the GF-expressions, however, with their values interpreted as integers.

The following example illustrates the first way of defining of the arithmetic expressions for multiple-valued functions by the example of ternary functions.

Example 24 *For single-variable ternary functions the basis functions used in the definition of the arithmetic expressions are*

$$\mathbf{X}_{3A}(1) = \begin{bmatrix} x^0 & x^1 & x^2 \end{bmatrix},$$

or in the explicit form as

$$\mathbf{X}_{3A}(1) = \begin{bmatrix} 1 & 0 & 0 \\ 1 & 1 & 1 \\ 1 & 2 & 4 \end{bmatrix}.$$

The arithmetic transform matrix used to calculate the coefficients in the arithmetic expression is

$$\mathbf{A}_3(1) = \mathbf{X}_3(1)^{-1} = \frac{1}{2} \begin{bmatrix} 2 & 0 & 0 \\ -3 & 4 & -1 \\ 1 & -2 & 1 \end{bmatrix}.$$

The extension to functions with an arbitrary number of variables is done through the Kronecker product of the basic transform matrix.

Example 25 *Consider the function f specified by the function vector $\mathbf{F} = [1, 1, 2, 2, 0, 1, 2, 1, 0]^T$. The arithmetic spectrum with respect to the basis $\mathbf{X}_{3A}(2)$ is $\mathbf{S}_{f,2A} = \frac{1}{4}[4, -2, 2, 6, -23, 9, -2, 11, -5]^T$.*

Example 26 *Consider the function f specified by the function vector $\mathbf{F} = [0, 1, 2, 1, 2, 0, 2, 0, 1]^T$. The arithmetic spectrum with respect to the basis $\mathbf{X}_{3A}(2)$ is $\mathbf{S}_{f,3A} = \frac{1}{4}[0, 4, 0, 4, 21, -15, 0, -15, 9]^T$.*

Example 27 *For $p = 4$, the arithmetic expressions are defined in terms of the basis functions specified by columns of the matrix*

$$\mathbf{X}_{4A}(1) = \begin{bmatrix} 1 & 0 & 0 & 0 \\ 1 & 1 & 1 & 1 \\ 1 & 2 & 4 & 8 \\ 1 & 3 & 9 & 27 \end{bmatrix}.$$

The matrix to determine coefficients in this arithmetic expressions is

$$\mathbf{A}_4(1) = \mathbf{X}_{4A}^{-1}(1) = \frac{1}{6} \begin{bmatrix} 6 & 0 & 0 & 0 \\ -11 & 18 & -9 & 2 \\ 6 & -15 & 12 & -3 \\ -1 & 3 & -3 & 1 \end{bmatrix}.$$

A problem with this definition of the arithmetic transform is that values which basis functions take are large, especially for a large value of p, since the exponent of x^{p-1} is taken. For example, if $p = 4$ and $n = 2$, the largest element of the corresponding matrix defining the basis functions is 729.

This bottleneck can be overcome if we keep the same set of basis functions as in the GF-expressions and interpret their values as integers instead of values in $GF(p)$. Then, we have the same set of basis functions as in the GF-expressions, however, under different interpretation of function values, the coefficients are integers which when scaled by a normalization factor can be used to represent multi-output functions in multiple-valued variables.

Example 28 *If the basis functions $\mathbf{X}_{3GF}(1)$, are interpreted as functions taking the corresponding integer values, the matrix inverse over the field of rational numbers Q defines the basic ternary arithmetic transform corresponding to the GF-expressions. This matrix is given by*

$$\mathbf{A}_3(1) = \mathbf{X}_3^{-1}(1) = \frac{1}{2} \begin{bmatrix} 2 & 0 & 0 \\ 0 & -2 & 2 \\ -2 & 4 & -2 \end{bmatrix}.$$

Again, extension to functions of an arbitrary number of variables is done through the Kronecker product of this basic transform matrix.

Example 29 *For the function* f *specified by the function vector* $\mathbf{F} = [1, 1, 2, 2, 0, 1, 2, 1, 0]^T$, *the arithmetic spectrum with respect to the basis* $\mathbf{X}_{3GF}$ *considered over* Q *is* $\mathbf{S}_{f,3A} = [1, 1, -1, 0, -2, 3, 1, 2, -5]^T$.

The optimization of these both classes of arithmetic expressions can be done by selecting polarities for variables in the same way as in the Galois field expressions for binary and multiple-valued cases. The restrictions to use p complements of variables remain valid in this case also due to the structure of the transform matrix.

2.7.2   ARITHMETIC EXPRESSIONS DERIVED FROM THE RMF-EXPRESSIONS

A disadvantage of GF-expressions is that except for $GF(2)$, the matrix expressing basis functions, and consequently the matrix used to calculate the coefficients in the expressions, is not triangular. As noticed above, this restricts the possible permutations which can be used in the corresponding fixed-polarity expressions. Thus, that approach towards generalizations of Reed-Muller expressions to multiple-valued logic functions has this property as a limitation in the optimization of the expressions by selecting polarities of variables. At the same time, the structure of the transform matrices reflects on the properties of the related expressions in the same ways as in the binary case and their resemblance or better to say discrepancy to the properties of the classical Fourier transform. This was a motivation for generalizations of the Reed-Muller expressions derived by their interpretation presented in [68]. In this way, the Reed-Muller-Fourier expressions have been defined.

The same considerations are true for the arithmetic equivalents of the RMF-expressions that are defined by using the same set of basis functions, however, interpreted as integer-valued functions, and with all the computations over the field of rational numbers Q.

Example 30 *For single-variable ternary functions, the basis to define the arithmetic expressions corresponding to the RMF-expressions is*

$$\mathbf{X}_{3ARMF}(1) = \begin{bmatrix} 2 & 0 & 0 \\ 2 & 1 & 0 \\ 2 & 2 & 2 \end{bmatrix}.$$

The coefficients in this expression are calculated by the arithmetic transform matrix defined as the inverse of $\mathbf{X}_3(1)$,

$$\mathbf{A}_{3RMF}(1) = \mathbf{X}_{3ARMF}^{-1}(1) = \frac{1}{2} \begin{bmatrix} 1 & 0 & 0 \\ -2 & 2 & 0 \\ 1 & -2 & 1 \end{bmatrix}.$$

Example 31 *As an illustration, consider the RMF-expression of a function* f *defined by the function vector* $\mathbf{F} = [1, 1, 2, 2, 0, 1, 2, 1, 0]^T$, *the RMF-coefficients are given by the vector* $\mathbf{S}_{f,3RMF} = [1, 0, 1, 2, 1, 1, 2, 0, 1]^T$, *and the RMF-expression is*

$$f = 1 \oplus x_2^{*2} \oplus 2x_1^{*1} \oplus (x_1^{*1} \odot x_2^{*1}) \oplus (x_1^{*1} \odot x_2^{*2}) \oplus 2x_1^{*2} \oplus (x_1^{*2} \odot x_2^{*2}).$$

The basis functions are identical to these used in the RMF-expressions, thus take the same values, however, interpreted as integers. Therefore, the basis functions written as columns of a (9×9) matrix are

$$\mathbf{X}_{3ARMF}(2) = \begin{bmatrix} 1 & 0 & 0 & 0 & 0 & 0 & 0 & 0 & 0 \\ 1 & 2 & 0 & 0 & 0 & 0 & 0 & 0 & 0 \\ 1 & 1 & 1 & 0 & 0 & 0 & 0 & 0 & 0 \\ 1 & 0 & 0 & 2 & 0 & 0 & 0 & 0 & 0 \\ 1 & 2 & 0 & 2 & 1 & 0 & 0 & 0 & 0 \\ 1 & 1 & 1 & 2 & 2 & 2 & 0 & 0 & 0 \\ 1 & 0 & 0 & 1 & 0 & 0 & 1 & 0 & 0 \\ 1 & 2 & 0 & 1 & 2 & 0 & 1 & 2 & 0 \\ 1 & 1 & 1 & 1 & 1 & 1 & 1 & 1 & 1 \end{bmatrix}.$$

The arithmetic spectrum corresponding to the RMF-expressions is defined by a matrix inverse to $\mathbf{X}_{3ARMF}(2)$,

$$\mathbf{A}_{3ARMF}(2) = \frac{1}{2} \begin{bmatrix} 2 & 0 & 0 & 0 & 0 & 0 & 0 & 0 & 0 \\ -1 & 1 & 0 & 0 & 0 & 0 & 0 & 0 & 0 \\ -1 & -1 & 2 & 0 & 0 & 0 & 0 & 0 & 0 \\ -1 & 0 & 0 & 1 & 0 & 0 & 0 & 0 & 0 \\ 2 & -2 & 0 & -2 & 2 & 0 & 0 & 0 & 0 \\ -1 & 2 & -1 & 1 & -2 & 1 & 0 & 0 & 0 \\ -1 & 0 & 0 & -1 & 0 & 0 & 2 & 0 & 0 \\ -1 & 1 & 0 & 2 & -2 & 0 & -1 & 1 & 0 \\ 2 & -1 & -1 & -1 & 2 & -1 & -1 & -1 & 2 \end{bmatrix}.$$

Thus, for the considered function f, the spectrum is computed as

$$\begin{aligned} \mathbf{S}_{f,3ARMF} &= \mathbf{A}_{3ARMF}(2)\mathbf{F} \\ &= \frac{1}{2}[2, 0, 2, 1, -4, 2, 1, 3, -7]^T. \end{aligned}$$

Thus, the arithmetic expression for the considered function f is

$$f = \frac{1}{2}(2 + 2x_2^{*2} + x_1 - 4(x_1 \odot x_2) + 2(x_1 \odot x_2^{*2}) + x_1^{*2} + 3(x_1^{*2} \odot x_2) - 7(x_1^{*2} \odot x_2^{*2})).$$

Example 32 *Consider the function f specified by the function vector $\mathbf{F} = [0, 1, 2, 1, 2, 0, 2, 0, 1]^T$. The arithmetic spectrum with respect to the basis $\mathbf{X}_{3ARMF}$ is $\mathbf{S}_{f,3ARMF} = \frac{1}{2}[0, 1, 3, 1, 0, -3, 3, -3, 0]^T$.*

The definition of RMF-expressions can be uniformly extended to functions for p non-prime. This will be illustrated for the example of $p = 4$.

The arithmetic expressions corresponding to the Reed-Muller-Fourier expressions for quaternary functions are defined in terms of this set of basis functions as

$$f(x_1, \ldots, x_n) = (-1)^n \mathbf{X}_{4ARMF}(n) \mathbf{S}_{f,4ARMF}(n),$$

where calculations are performed in Q.

Notice that both the matrices defining basis functions and transform matrices used to calculate the coefficients are triangular matrices with upper-right part consisting of zero elements. Due to this, it is possible to exploit all $p!$ permutations of a p-valued variable as its complements. In this way, the number of different expressions for a function of n variables is extended from p^n into $(p!)^n$, which increases possibilities to determine expressions with reduced number of non-zero coefficients as compared to Galois field expressions. At the same time, all properties corresponding to properties of the Fourier representations for the binary case are preserved.

2.7.3 STRUCTURE OF THE ARITHMETIC RMF-TRANSFORM MATRICES

To extend applicability of RMF-expressions to integer-valued functions, the arithmetic RMF-expressions are defined by using the same basis functions, however, with their values $\{0, 1, \ldots, p - 1\}$ interpreted as integers. Then, we calculate the inverse of the matrix $\mathbf{X}_{pRMF}(n)$ over the field of rational numbers Q. This matrix cannot be represented by the Kronecker product, however, it possesses a recursive structure as will be illustrated below by the example for functions in ternary variables [204], [207].

First, we define $\mathbf{A}_3(0) = 2$, $\mathbf{B}_3(0) = 1$, and $\mathbf{C}_3(0) = 1$. Then,

$$\mathbf{A}_{3,ARMF}(n) = \begin{bmatrix} \mathbf{B}_3(n-1) & \mathbf{0}_3(n-1) & \mathbf{0}_3(n-1) \\ -\mathbf{A}_3(n-1) & \mathbf{A}_3(n-1) & \mathbf{0}_3(n-1) \\ \mathbf{C}_3(n-1) & -\mathbf{A}_3(n-1) & \mathbf{B}_3(n-1) \end{bmatrix},$$

$$\mathbf{B}_{3,ARMF}(n) = \begin{bmatrix} \mathbf{A}_3(n-1) & \mathbf{0}_3(n-1) & \mathbf{0}_3(n-1) \\ -\mathbf{B}_3(n-1) & \mathbf{B}_3(n-1) & \mathbf{0}_3(n-1) \\ -\mathbf{C}_3(n-1) & -\mathbf{B}_3(n-1) & \mathbf{A}_3(n-1) \end{bmatrix},$$

$$\mathbf{C}_{3,ARMF}(n) = \begin{bmatrix} -\mathbf{C}_3(n-1) & \mathbf{0}_3(n-1) & \mathbf{0}_3(n-1) \\ -\mathbf{C}_3(n-1) & \mathbf{C}_3(n-1) & \mathbf{0}_3(n-1) \\ 2\mathbf{C}_3(n-1) & -\mathbf{C}_3(n-1) & -\mathbf{C}_3(n-1) \end{bmatrix}.$$

Notice that the normalization factor in this Arithmetic RMF-transform is $1/2$.

Example 33 *For* $n = 1$, *the Arithmetic RMF-transform matrix is computed as the inverse of* $\mathbf{X}_{3,ARMF}(1) = \begin{bmatrix} 2 & 0 & 0 \\ 2 & 1 & 0 \\ 2 & 2 & 2 \end{bmatrix}$, *where* $\mathbf{A}_{3,ARMF}(1) = \frac{1}{2} \begin{bmatrix} 1 & 0 & 0 \\ -2 & 2 & 0 \\ 1 & -2 & 1 \end{bmatrix}$ *is the inverse of* $\mathbf{X}_{3,ARMF}(1)$.

The matrices $\mathbf{B}_3(1)$ *and* $\mathbf{C}_3(1)$ *are*

$$\mathbf{B}_{3,ARMF}(1) = \begin{bmatrix} 2 & 0 & 0 \\ -1 & 1 & 0 \\ 1 & -1 & 2 \end{bmatrix}, \qquad \mathbf{C}_{3,ARMF}(1) = \begin{bmatrix} -1 & 0 & 0 \\ -1 & 1 & 0 \\ 2 & -1 & -1 \end{bmatrix}.$$

For $n = 2$, *the matrix* $\mathbf{A}_{3,ARMF}(2)$ *is the inverse of* $\mathbf{X}_{3,ARMF}(2)$ *and it determined as*

$$
\mathbf{A}_{3,ARMF}(2) = \frac{1}{2}\begin{bmatrix} \mathbf{B}_3(1) & \mathbf{0}_3(1) & \mathbf{0}_3(1) \\ -\mathbf{A}_3(1) & \mathbf{A}_3(1) & \mathbf{0}_3(1) \\ \mathbf{C}_3(1) & -\mathbf{A}_3(1) & \mathbf{B}_3(1) \end{bmatrix}
$$

$$
= \frac{1}{2}\left[\begin{array}{ccc|ccc|ccc}
2 & 0 & 0 & 0 & 0 & 0 & 0 & 0 & 0 \\
-1 & 1 & 0 & 0 & 0 & 0 & 0 & 0 & 0 \\
-1 & -1 & 2 & 0 & 0 & 0 & 0 & 0 & 0 \\
-1 & 0 & 0 & 1 & 0 & 0 & 0 & 0 & 0 \\
2 & -2 & 0 & -2 & 2 & 0 & 0 & 0 & 0 \\
-1 & 2 & -1 & 1 & -2 & 1 & 0 & 0 & 0 \\
-1 & 0 & 0 & -1 & 0 & 0 & 2 & 0 & 0 \\
-1 & 1 & 0 & 2 & -2 & 0 & -1 & 1 & 0 \\
2 & -1 & -1 & -1 & 2 & -1 & -1 & -1 & 2
\end{array}\right].
$$

In a similar way,

$$
\mathbf{B}_{3,ARMF}(2) = \begin{bmatrix} \mathbf{A}_3(1) & \mathbf{0}_3(1) & \mathbf{0}_3(1) \\ -\mathbf{B}_3(1) & \mathbf{B}_3(1) & \mathbf{0}_3(1) \\ -\mathbf{C}_3(1) & -\mathbf{B}_3(1) & \mathbf{A}_3(1) \end{bmatrix}
$$

$$
= \frac{1}{2}\left[\begin{array}{ccc|ccc|ccc}
1 & 0 & 0 & 0 & 0 & 0 & 0 & 0 & 0 \\
-2 & 2 & 0 & 0 & 0 & 0 & 0 & 0 & 0 \\
1 & -2 & 1 & 0 & 0 & 0 & 0 & 0 & 0 \\
-2 & 0 & 0 & 2 & 0 & 0 & 0 & 0 & 0 \\
1 & -1 & 0 & -1 & 1 & 0 & 0 & 0 & 0 \\
1 & 1 & -2 & -1 & -1 & 2 & 0 & 0 & 0 \\
1 & 0 & 0 & -2 & 0 & 0 & 1 & 0 & 0 \\
1 & -1 & 0 & 1 & -1 & 0 & -2 & 2 & 0 \\
-2 & 1 & 1 & 1 & 1 & -2 & -1 & -2 & 1
\end{array}\right],
$$

and

$$
\mathbf{C}_{3,ARMF}(2) = \begin{bmatrix} -\mathbf{C}_3(1) & \mathbf{0}_3(1) & \mathbf{0}_3(1) \\ -\mathbf{C}_3(1) & \mathbf{C}_3(1) & \mathbf{0}_3(1) \\ 2\mathbf{C}_3(1) & -\mathbf{C}_3(1) & -\mathbf{C}_3(1) \end{bmatrix}
$$

$$
= \frac{1}{2} \left[\begin{array}{rrr|rrr|rrr} 1 & 0 & 0 & 0 & 0 & 0 & 0 & 0 & 0 \\ 1 & -1 & 0 & 0 & 0 & 0 & 0 & 0 & 0 \\ -2 & 1 & 1 & 0 & 0 & 0 & 0 & 0 & 0 \\ \hline 1 & 0 & 0 & -1 & 0 & 0 & 0 & 0 & 0 \\ 1 & -1 & 0 & -1 & 1 & 0 & 0 & 0 & 0 \\ -2 & 1 & 1 & 2 & -1 & -1 & 0 & 0 & 0 \\ \hline -2 & 0 & 0 & 1 & 0 & 0 & 1 & 0 & 0 \\ -2 & 2 & 0 & 1 & -1 & 0 & 1 & -1 & 0 \\ 4 & -2 & -2 & -2 & 1 & 1 & -2 & 1 & 1 \end{array} \right].
$$

The matrix $\mathbf{A}_3(n)$ is used to calculate coefficients in the arithmetic RMF-expressions for ternary functions defined as

$$
f(x_1, \ldots, x_n) = \sum_{i=0}^{3^n-1} a_i s(i),
$$

where $s(i)$ are columns of the matrix $\mathbf{X}_{3RMF}(n)$, with entries interpreted as integers. In other words, $s(i)$ are $3AND$ product of ternary variables to the integer powers in terms of $3EXP$ in Hadamard ordering.

In general, we may conclude that word-level expressions for multiple-valued functions which we considered are defined in two ways:

1. by assigning to each p-valued variable a polynomial of order p in terms of exponentiation as an operation in the field of rational numbers Q, and using these polynomials to define the set of basis functions for n-variable functions;

2. by preserving the same set of basis functions as in modulo p structures, however, with interpretation of function values in the set of integers, to determine coefficients in the expressions considered.

For further discussions of the arithmetic representations for multiple-valued functions, we refer to [10], [236], [237], [249], [250], [251].

For Reed-Muller-like expressions of representation of incompletely specified multiple-valued functions; see for example [79], [80], [96], [97], [258], [259], [262], [264], [265], [266], [267].

Various other generalizations in terms of combinations of either modular operations, or *min* and *max* operations, and literal operators, are not discussed. In a way, the expressions based on *min* and *max* operations (as a generalization of logic AND and OR), as well as literal operators, can be rather viewed as generalizations of Sum-of-Product (SOP) expressions and related representations for two-valued (Boolean) logic functions, than as a generalization of polynomial or spectral representations. For these representations we refer to [39], [40], [48], [102]. There are also various combinations of different functional expressions aimed at compact representations, for instance, see [122], [123], [211].

For optimization methods we refer to [27], [55], [56], [57], [58], [59], [60], [61], [62], [60], [63], [64], [65], [70], [71], [81], [121], [156], [165], [169], [172], [173], [174], [175], [176], [253], [255]. An approach to the minimization of Galois field expressions through a generalization of the notion of special normal forms in Boolean functions [223] to ternary logic functions has been presented in [193].

We also did not discuss representations where the main intention was to reduce the number of non-zero coefficients while preserving fast calculation algorithms, however, at the price of discharging other properties usually expressed by spectral (Fourier series-like) expressions. For such generalizations of Reed-Muller expressions we refer, for instance, to [33], [54], [56], [57], [59], [65].

2.8   HAAR-LIKE EXPRESSIONS FOR MULTIPLE-VALUED FUNCTIONS

In the binary case, Haar functions and Haar transform differ from the Walsh functions and the Walsh transform by the property that they are not Kronecker product representable, but layer-Kronecker representable; for instance, see [50], [51], [52]. Since at each layer the submatrices used in the Kronecker products to generate the Haar matrix are either the basic Walsh matrix or some of its rows or columns (we use rows or columns, depending on whether we consider the matrices for the direct transforms or for the inverse transforms), there is a resemblance between these function systems and transforms and their properties.

This structure of the matrix defining Haar functions $\mathbf{H}(n)$, and correspondingly, the Haar transform matrix $\mathbf{H}^{-1}(n) = \mathbf{H}^T(n)$ (due to the orthogonality [74]), allows for various generalizations by appropriately selecting different basic transform matrices in the Kronecker products at different layers of the matrices. This approach has been discussed in [220] for the case of using Fourier transforms and Galois field transforms. This approach will be briefly discussed and a bit further elaborated here by the example of Haar expressions related to the RMF-expressions.

The Haar-RMF-transform for $n > 1$ is defined by the transform matrix

$$\mathbf{R}_h(n) = \left[\begin{array}{c} \mathbf{R}_h(n-1) \otimes \mathbf{r}_0^T \\ \mathbf{I}(n-1) \otimes \mathbf{r}_1^T \\ \mathbf{I}(n-1) \otimes \mathbf{r}_2^T \\ \mathbf{I}(n-1) \otimes \mathbf{r}_3^T \end{array} \right],$$

where $\mathbf{r}_i, i = 0, 1, 2, 3$, are columns of the basic RMF-matrix $\mathbf{R}_{RMF}(1)$, and $\mathbf{I}(n-1)$ is the $((n-1) \times (n-1))$ identity matrix. For $n = 1, \mathbf{R}_h(1) = \mathbf{R}_{4RMF}(1)$.

The inverse transform, i.e., the matrix whose columns are the set of basis functions in terms of which this transform is defined is

$$\mathbf{X}_h(n) = \left[\ \mathbf{X}_h(n-1) \otimes [1,0,0,0]^T,\quad \mathbf{I}(n-1) \otimes 3\mathbf{x}_1^T,\quad \mathbf{I}(n-1) \otimes 3\mathbf{x}_2^T,\quad \mathbf{I}(n-1) \otimes 3\mathbf{x}_3^T\ \right],$$

where $\mathbf{x}_i$, $i = 1, 2, 3$, are rows of $\mathbf{X}_{4RMF}(1)$. For $n = 1$, $\mathbf{X}_h(1) = \mathbf{X}_{4RMF}(1)$.

Example 34 *For $p = 4$ and $n = 2$, the Haar-RMF-transform is defined by the following transform matrix. Recall that*

$$\mathbf{R}_{4RMF}(1) = 3 \begin{bmatrix} 1 & 0 & 0 & 0 \\ 1 & 3 & 0 & 0 \\ 1 & 2 & 1 & 0 \\ 1 & 1 & 3 & 3 \end{bmatrix}.$$

Then,

$$\mathbf{R}_h(2) = \begin{bmatrix} \mathbf{R}_h(1) \otimes [3,3,3,3] \\ \mathbf{I}(1) \otimes [0,1,2,3] \\ \mathbf{I}(1) \otimes [0,0,3,1] \\ \mathbf{I}(1) \otimes [0,0,0,1] \end{bmatrix}$$

$$= \left[\begin{array}{cccc|cccc|cccc|cccc}
1 & 1 & 1 & 1 & 0 & 0 & 0 & 0 & 0 & 0 & 0 & 0 & 0 & 0 & 0 & 0 \\
1 & 1 & 1 & 1 & 3 & 3 & 3 & 3 & 0 & 0 & 0 & 0 & 0 & 0 & 0 & 0 \\
1 & 1 & 1 & 1 & 2 & 2 & 2 & 2 & 1 & 1 & 1 & 1 & 0 & 0 & 0 & 0 \\
1 & 1 & 1 & 1 & 1 & 1 & 1 & 1 & 3 & 3 & 3 & 3 & 3 & 3 & 3 & 3 \\
\hline
0 & 1 & 2 & 3 & 0 & 0 & 0 & 0 & 0 & 0 & 0 & 0 & 0 & 0 & 0 & 0 \\
0 & 0 & 0 & 0 & 0 & 1 & 2 & 3 & 0 & 0 & 0 & 0 & 0 & 0 & 0 & 0 \\
0 & 0 & 0 & 0 & 0 & 0 & 0 & 0 & 0 & 1 & 2 & 3 & 0 & 0 & 0 & 0 \\
0 & 0 & 0 & 0 & 0 & 0 & 0 & 0 & 0 & 0 & 0 & 0 & 0 & 1 & 2 & 3 \\
\hline
0 & 0 & 3 & 1 & 0 & 0 & 0 & 0 & 0 & 0 & 0 & 0 & 0 & 0 & 0 & 0 \\
0 & 0 & 0 & 0 & 0 & 0 & 3 & 1 & 0 & 0 & 0 & 0 & 0 & 0 & 0 & 0 \\
0 & 0 & 0 & 0 & 0 & 0 & 0 & 0 & 0 & 0 & 3 & 1 & 0 & 0 & 0 & 0 \\
0 & 0 & 0 & 0 & 0 & 0 & 0 & 0 & 0 & 0 & 0 & 0 & 0 & 0 & 3 & 1 \\
\hline
0 & 0 & 0 & 1 & 0 & 0 & 0 & 0 & 0 & 0 & 0 & 0 & 0 & 0 & 0 & 0 \\
0 & 0 & 0 & 0 & 0 & 0 & 0 & 1 & 0 & 0 & 0 & 0 & 0 & 0 & 0 & 0 \\
0 & 0 & 0 & 0 & 0 & 0 & 0 & 0 & 0 & 0 & 0 & 1 & 0 & 0 & 0 & 0 \\
0 & 0 & 0 & 0 & 0 & 0 & 0 & 0 & 0 & 0 & 0 & 0 & 0 & 0 & 0 & 1
\end{array}\right].$$

The set of basis functions in terms of which this transform is defined, when written as columns of the matrix inverse to $\mathbf{R}_h(2)$, *are specified as*

$$
\mathbf{X}_h(2) = \left[
\begin{array}{cccc|cccc|cccc|cccc}
3 & 0 & 0 & 0 & 1 & 0 & 0 & 0 & 1 & 0 & 0 & 0 & 1 & 0 & 0 & 0 \\
0 & 0 & 0 & 0 & 3 & 0 & 0 & 0 & 2 & 0 & 0 & 0 & 1 & 0 & 0 & 0 \\
0 & 0 & 0 & 0 & 0 & 0 & 0 & 0 & 1 & 0 & 0 & 0 & 3 & 0 & 0 & 0 \\
0 & 0 & 0 & 0 & 0 & 0 & 0 & 0 & 0 & 0 & 0 & 0 & 3 & 0 & 0 & 0 \\
\hline
3 & 1 & 0 & 0 & 0 & 1 & 0 & 0 & 0 & 1 & 0 & 0 & 0 & 1 & 0 & 0 \\
0 & 0 & 0 & 0 & 0 & 3 & 0 & 0 & 0 & 2 & 0 & 0 & 0 & 1 & 0 & 0 \\
0 & 0 & 0 & 0 & 0 & 0 & 0 & 0 & 0 & 1 & 0 & 0 & 0 & 3 & 0 & 0 \\
0 & 0 & 0 & 0 & 0 & 0 & 0 & 0 & 0 & 0 & 0 & 0 & 0 & 3 & 0 & 0 \\
\hline
3 & 2 & 3 & 0 & 0 & 0 & 1 & 0 & 0 & 0 & 1 & 0 & 0 & 0 & 1 & 0 \\
0 & 0 & 0 & 0 & 0 & 0 & 3 & 0 & 0 & 0 & 2 & 0 & 0 & 0 & 1 & 0 \\
0 & 0 & 0 & 0 & 0 & 0 & 0 & 0 & 0 & 0 & 1 & 0 & 0 & 0 & 3 & 0 \\
0 & 0 & 0 & 0 & 0 & 0 & 0 & 0 & 0 & 0 & 0 & 0 & 0 & 0 & 3 & 0 \\
\hline
3 & 3 & 1 & 1 & 0 & 0 & 0 & 1 & 0 & 0 & 0 & 1 & 0 & 0 & 0 & 1 \\
0 & 0 & 0 & 0 & 0 & 0 & 0 & 3 & 0 & 0 & 0 & 2 & 0 & 0 & 0 & 1 \\
0 & 0 & 0 & 0 & 0 & 0 & 0 & 0 & 0 & 0 & 0 & 1 & 0 & 0 & 0 & 3 \\
0 & 0 & 0 & 0 & 0 & 0 & 0 & 0 & 0 & 0 & 0 & 0 & 0 & 0 & 0 & 3 \\
\end{array}
\right].
$$

Example 35 *For the function* f *specified by the function vector* $\mathbf{F} = [2, 0, 0, 1, 3, 3, 3, 2, 3, 3, 3, 0, 0, 1, 1, 2]^T$, *the Haar-RMF-spectrum is computed as*

$$
\mathbf{S}_{f, h4RMF} = 3\mathbf{R}_h(2)\mathbf{F},
$$

and it is $\mathbf{S}_{f, h4RMF} = [1, 0, 2, 3, 1, 1, 3, 3, 3, 1, 3, 3, 3, 2, 0, 2]^T$.

An alternative way to define the Haar-like RMF-transform, is illustrated by the following example. The Haar-RMF matrices defined as above are used in Chapter 4 to define decision diagrams that belong to the class of Haar Spectral Transform Decision Diagrams (HSTDDs); for instance, see [199].

The set of basis functions for the Haar-RMF-expressions for quaternary functions is defined as

$$
\mathbf{X}_{h, 4RMF}(n) = \left[
\begin{array}{c}
\mathbf{r}_0 \otimes \mathbf{R}_h(n-1) \\
3\mathbf{r}_1 \otimes \mathbf{I}(n-1) \\
3\mathbf{r}_2 \otimes \mathbf{I}(n-1) \\
3\mathbf{r}_3 \otimes \mathbf{I}(n-1)
\end{array}
\right],
$$

where $\mathbf{r}_i, i = 0, 1, 2, 3$, are columns of $\mathbf{R}(1)$, and $\mathbf{I}(n-1)$ is the $((n-1) \times (n-1))$ identity matrix.

The corresponding transform matrix that is used to determine the Haar-RMF coefficients is defined as

$$\mathbf{Y}_{h,4RMF}(n) = \begin{bmatrix} \mathbf{q}_0 \otimes \mathbf{R}_h(n-1) \\ 3\mathbf{q}_1 \otimes \mathbf{I}(n-1) \\ 3\mathbf{q}_2 \otimes \mathbf{I}(n-1) \\ 3\mathbf{q}_3 \otimes \mathbf{I}(n-1) \end{bmatrix},$$

where $\mathbf{q}_i$, $i = 0, 1, 2, 3$, are rows of $\mathbf{R}(1)$.

Example 36 *The set of basis functions in terms of which the Haar-RMF-expressions for quaternary functions of two variables are defined are specified by columns of a matrix determined as*

$$\mathbf{X}_{yh,4RMF}(2) = \begin{bmatrix} \mathbf{r}_0 \otimes \mathbf{R}(1) & 3\mathbf{r}_1 \otimes \mathbf{I}(1) & 3\mathbf{r}_2 \otimes \mathbf{I}(1) & 3\mathbf{r}_3 \otimes \mathbf{I}(1) \end{bmatrix}$$

$$= \begin{bmatrix} \begin{bmatrix} 3 \\ 3 \\ 3 \\ 3 \end{bmatrix} \otimes \mathbf{R}(1) & \begin{bmatrix} 0 \\ 1 \\ 2 \\ 3 \end{bmatrix} \otimes \mathbf{I}(1) & \begin{bmatrix} 0 \\ 0 \\ 3 \\ 1 \end{bmatrix} \otimes \mathbf{I}(1) & \begin{bmatrix} 0 \\ 0 \\ 0 \\ 1 \end{bmatrix} \otimes \mathbf{I}(1) \end{bmatrix}$$

$$= \begin{bmatrix}
1 & 0 & 0 & 0 & 0 & 0 & 0 & 0 & 0 & 0 & 0 & 0 & 0 & 0 & 0 & 0 \\
1 & 3 & 0 & 0 & 0 & 0 & 0 & 0 & 0 & 0 & 0 & 0 & 0 & 0 & 0 & 0 \\
1 & 2 & 1 & 0 & 0 & 0 & 0 & 0 & 0 & 0 & 0 & 0 & 0 & 0 & 0 & 0 \\
1 & 1 & 3 & 3 & 0 & 0 & 0 & 0 & 0 & 0 & 0 & 0 & 0 & 0 & 0 & 0 \\
1 & 0 & 0 & 0 & 3 & 0 & 0 & 0 & 0 & 0 & 0 & 0 & 0 & 0 & 0 & 0 \\
1 & 3 & 0 & 0 & 0 & 3 & 0 & 0 & 0 & 0 & 0 & 0 & 0 & 0 & 0 & 0 \\
1 & 2 & 1 & 0 & 0 & 0 & 3 & 0 & 0 & 0 & 0 & 0 & 0 & 0 & 0 & 0 \\
1 & 1 & 3 & 3 & 0 & 0 & 0 & 3 & 0 & 0 & 0 & 0 & 0 & 0 & 0 & 0 \\
1 & 0 & 0 & 0 & 2 & 0 & 0 & 0 & 1 & 0 & 0 & 0 & 0 & 0 & 0 & 0 \\
1 & 3 & 0 & 0 & 0 & 2 & 0 & 0 & 0 & 1 & 0 & 0 & 0 & 0 & 0 & 0 \\
1 & 2 & 1 & 0 & 0 & 0 & 2 & 0 & 0 & 0 & 1 & 0 & 0 & 0 & 0 & 0 \\
1 & 1 & 3 & 3 & 0 & 0 & 0 & 2 & 0 & 0 & 0 & 1 & 0 & 0 & 0 & 0 \\
1 & 0 & 0 & 0 & 1 & 0 & 0 & 0 & 3 & 0 & 0 & 0 & 3 & 0 & 0 & 0 \\
1 & 3 & 0 & 0 & 0 & 1 & 0 & 0 & 0 & 3 & 0 & 0 & 0 & 3 & 0 & 0 \\
1 & 2 & 1 & 0 & 0 & 0 & 1 & 0 & 0 & 0 & 3 & 0 & 0 & 0 & 3 & 0 \\
1 & 1 & 3 & 3 & 0 & 0 & 0 & 1 & 0 & 0 & 0 & 3 & 0 & 0 & 0 & 3
\end{bmatrix}.$$

These basis functions can be expressed as

$$\mathbf{X}_{yh,4RMF}(2) = [1, 3x_2, 3x_2^{*2}, 3x_2^{*3}, x_1 J_0(x_2), x_1 J_1(x_2), x_1 J_2(x_2), x_1 J_3(x_2),$$
$$x_1^{*2} J_0(x_2), x_1^{*2} J_1(x_2), x_1^{*2} J_2(x_2), x_1^{*2} J_3(x_2), \qquad (2.12)$$
$$x_1^{*3} J_0(x_2), x_1^{*3} J_1(x_2), x_1^{*3} J_2(x_2), x_1^{*3} J_3(x_2)]^T.$$

The Haar-RMF matrix is defined as the inverse of $\mathbf{X}_{yh,4RMF}(2)$ *and is given by*

$$
\mathbf{Y}_{h,4RMF}(2) \;=\;
\begin{bmatrix}
\mathbf{q}_0 \otimes \mathbf{R}(1) \\
3\mathbf{q}_1 \otimes \mathbf{I}(1) \\
3\mathbf{q}_2 \otimes \mathbf{I}(1) \\
3\mathbf{q}_3 \otimes \mathbf{I}(1)
\end{bmatrix}
\tag{2.13}
$$

$$
=\;
\begin{bmatrix}
\begin{bmatrix} 3 & 0 & 0 & 0 \end{bmatrix} \otimes \mathbf{R}(1) \\
\begin{bmatrix} 1 & 3 & 0 & 0 \end{bmatrix} \otimes \mathbf{I}(1) \\
\begin{bmatrix} 1 & 2 & 1 & 0 \end{bmatrix} \otimes \mathbf{I}(1) \\
\begin{bmatrix} 1 & 1 & 3 & 3 \end{bmatrix} \otimes \mathbf{I}(1)
\end{bmatrix}
\tag{2.14}
$$

$$
=\;
\left[
\begin{array}{cccc|cccc|cccc|cccc}
1 & 0 & 0 & 0 & 0 & 0 & 0 & 0 & 0 & 0 & 0 & 0 & 0 & 0 & 0 & 0 \\
1 & 3 & 0 & 0 & 0 & 0 & 0 & 0 & 0 & 0 & 0 & 0 & 0 & 0 & 0 & 0 \\
1 & 2 & 1 & 0 & 0 & 0 & 0 & 0 & 0 & 0 & 0 & 0 & 0 & 0 & 0 & 0 \\
1 & 1 & 3 & 3 & 0 & 0 & 0 & 0 & 0 & 0 & 0 & 0 & 0 & 0 & 0 & 0 \\
\hline
1 & 0 & 0 & 0 & 3 & 0 & 0 & 0 & 0 & 0 & 0 & 0 & 0 & 0 & 0 & 0 \\
0 & 1 & 0 & 0 & 0 & 3 & 0 & 0 & 0 & 0 & 0 & 0 & 0 & 0 & 0 & 0 \\
0 & 0 & 1 & 0 & 0 & 0 & 3 & 0 & 0 & 0 & 0 & 0 & 0 & 0 & 0 & 0 \\
0 & 0 & 0 & 1 & 0 & 0 & 0 & 3 & 0 & 0 & 0 & 0 & 0 & 0 & 0 & 0 \\
\hline
1 & 0 & 0 & 0 & 2 & 0 & 0 & 0 & 1 & 0 & 0 & 0 & 0 & 0 & 0 & 0 \\
0 & 1 & 0 & 0 & 0 & 2 & 0 & 0 & 0 & 1 & 0 & 0 & 0 & 0 & 0 & 0 \\
0 & 0 & 1 & 0 & 0 & 0 & 2 & 0 & 0 & 0 & 1 & 0 & 0 & 0 & 0 & 0 \\
0 & 0 & 0 & 1 & 0 & 0 & 0 & 2 & 0 & 0 & 0 & 1 & 0 & 0 & 3 & 0 \\
\hline
1 & 0 & 0 & 0 & 1 & 0 & 0 & 0 & 3 & 0 & 0 & 0 & 3 & 0 & 0 & 0 \\
0 & 1 & 0 & 0 & 0 & 1 & 0 & 0 & 0 & 3 & 0 & 0 & 0 & 3 & 0 & 0 \\
0 & 0 & 1 & 0 & 0 & 0 & 1 & 0 & 0 & 0 & 3 & 0 & 0 & 0 & 3 & 0 \\
0 & 0 & 0 & 1 & 0 & 0 & 0 & 1 & 0 & 0 & 0 & 3 & 0 & 0 & 0 & 3
\end{array}
\right].
\tag{2.15}
$$

Example 37 For the function f specified by the function vector $\mathbf{F} = [2,0,0,1,3,3,3,2,3,3,3,0,0,1,1,2]^T$, the Haar-RMF-spectrum is $\mathbf{S}_{f,yh4RMD} = [2,2,2,1,1,3,3,1,1,3,3,3,2,1,1,3]^T$. For the function f whose truth-vector is $\mathbf{F} = [0,0,0,0,0,1,2,3,0,3,2,1,0,2,1,3]^T$, the Haar-RMF-spectrum is $\mathbf{S}_{f,yh4RMF} = [0,0,0,0,0,3,2,1,0,1,2,3,0,0,3,3]^T$.

2.9   SPARSE REPRESENTATIONS FROM COVERING CODES

In previous sections, we discuss the optimization methods for functional expressions by using the number of non-zero coefficients as the complexity measure. An approach is the following. For a given function, we select the basis functions that will provide small number of non-zero coefficients. The problem is that there are no proper guidelines which basis to select, except for some very specific classes of functions where the practice has shown that certain bases can be suitable. For instance, in the case of binary logic functions, the arithmetic functions as adders, multipliers, etc., can be compactly represented by functional expressions derived from the arithmetic spectral transforms [230]. The elementary mathematical functions as sine, cosine, sigmoid, logarithm, etc., can also be compactly represented by the arithmetic transform expressions which, therefore, were used as a basis for the design of the related hardware implementations [146], [147], [197].

These, however, are just few classes of functions for which such recommendations have been proposed. In the case of multiple-valued functions an even smaller amount of work has been done. Another approach is to select a given class of functional expressions, as for instance the Reed-Muller expressions and their generalizations to multiple-valued functions, and then perform the optimization by using different polarity for variables, or try to combine different expansion rules for variables. This later approach leads to different representations as Kronecker representations, or Pseudo-Kronecker representations, Generalized Reed-Muller expressions, etc. The main background idea is to increase the number of possible representations for a given function increasing in this way the possibility to find a representation with a small or an acceptable number of non-zero coefficients for the applications intended.

A considerable amount of work has been done by Professor Bogdan J. Falkowski and his associates in defining a variety of transforms with the main criteria to reduce the number of non-zero coefficients for representations of different classes of multiple-valued functions and to preserve the existence of fast FFT-like computation algorithms; for instance, see [54], [59], [63], [65], [154], [155], [157].

In either of these approaches, the number of non-zero coefficients required to represent a given function is not guaranteed in advance. It might happen that after an extensive search, the expressions with a given number of terms cannot be found. This bottleneck can be overcome by using redundant bases determined from error-correcting codes [13].

By borrowing some basic results from the *Theory of Error-Correcting Codes*, see for instance [110], we are able to construct a basis in terms of which a given function can be represented by no more than an a priori specified (small) number of non-zero coefficients. The decoding methods provide a way to determine these coefficients. A useful feature is that the basis functions are often easy to generate. Taking this as a motivation, in this section, we briefly discuss applications of covering codes in designing functional expressions with an a priori specified number of product terms. The presentation is mainly based on the work reported in [13].

2.9.1   COVERING CODES

In this section we review a few basic concepts of linear error-correcting block codes that will be used latter.

Consider a finite field $GF(q)$ with q elements. We define a code C as a subset of the linear space $GF(q^n)$ of q-ary sequences of length n. Now, the code C is called *a linear code* iff C is a linear subspace of $GF(q^n)$. It is clear that if C has dimension $k \leq n$, then it is spanned by k linearly independent vectors of C. The matrix $\mathbf{G}$ having as rows vectors of any such set is *a generator matrix* of the code C. A code of length n and dimension k is called an (n, k) *code*.

Let $\mathbf{G}'$ be a generator matrix of C. It is clear that permuting columns if necessary (which leads to an equivalent code) we can convert $\mathbf{G}'$ into the *systematic form*

$$
\mathbf{G} = [\mathbf{I}_k | \mathbf{P}]
$$
$$
= \begin{bmatrix}
1 & 0 & 0 & \cdots & 0 & p_{1,1} & p_{1,2} & \cdots & p_{1,n-k} \\
0 & 1 & 0 & \cdots & 0 & p_{2,1} & p_{2,2} & \cdots & p_{2,n-k} \\
& & & \vdots & & & & & \vdots \\
0 & 0 & 0 & \cdots & 1 & p_{k,1} & p_{k,2} & \cdots & p_{k,n-k}
\end{bmatrix}.
$$

Therefore, $\mathbf{G}$ is an $(n \times k)$ matrix, which consists of a $(k \times k)$ identity matrix and a $(k \times (n-k))$ matrix $\mathbf{P}$ which is the *parity* part of the matrix. These names reflect the fact that the code C transforms a length k information word $\mathbf{i} = [i_1, \ldots, i_k]$ to a length n codeword $\mathbf{c} = [c_1, \ldots, c_n]$ by the matrix multiplication

$$
\mathbf{c} = \mathbf{i}^T \cdot \mathbf{G} = [i_1, \ldots, i_k][\mathbf{I}_k | \mathbf{P}].
$$

Thus, in the codeword $\mathbf{c}$ the first k bits (digits for $GF(q)$) are *information bits* and the last $n-k$ are *parity check bits*.

We can equivalently specify the dimension k subspace C of $GF(q^n)$ by listing $n-k$ linearly independent vectors of $C^\perp$ (i.e., orthogonal to C). Any $((n-k) \times k)$ matrix with the vectors of such a set as rows is a *parity check matrix* of C. It is clear that if

$$
\mathbf{G} = [\mathbf{I}_k | \mathbf{P}],
$$

is a generator matrix of C, then the $((n-k) \times k)$ matrix

$$
\mathbf{H} = [-\mathbf{P}^T | \mathbf{I}_{n-k}]
$$

is a parity check matrix for C. If C is binary, naturally, $-\mathbf{P} = \mathbf{P}$.

The error-correcting and covering properties of a code are defined in terms of the *Hamming distance*.

Definition 17 *(The Hamming distance)*
The Hamming distance $d_H(x, y)$ of two vectors x and y of the same length is defined as the number of coordinates where x and y differ.

Definition 18 *(The Hamming weight)*
The Hamming weight $w(x)$ of a vector x is the number of non-zero elements in x.

Definition 19 *(Error-correcting code)*
A code is e-error-correcting if the minimum Hamming distance between codewords is $2e + 1'$.

Definition 20 *(Covering code)*
A code $C \in GF(q^n)$ is called a q-ary r-covering code of length n if for every word $y \in GF(q^n)$ there is a codeword $x \in C$ such that the Hamming distance $d_H(x, y) \leq r$. The smallest value for r satisfying this requirement is called the covering radius of the code. In other words, the finite metric space $GF(q^n)$ is exhausted by the spheres of radius r around the codewords of C.

Another important notion are *perfect codes*.

Definition 21 *(Perfect code)*
A code is called perfect if it is e-error-correcting and its covering radius is e.

For general properties of error-correcting and covering codes we refer to [110].

2.9.2   FUNCTIONAL EXPRESSIONS DETERMINED FROM COVERING CODES

Assume that an (n, k) linear code C has covering radius r. This is equivalent to the statement that any vector of length $n - k$ can be expressed as a sum of at most r columns of the parity check matrix

$$\mathbf{H} = [\mathbf{P}^T | \mathbf{I}_{n-k}]$$

$$= \begin{bmatrix} p_{1,1} & p_{1,2} & \cdots & p_{1,k} & 1 & 0 & \cdots & 0 \\ p_{2,1} & p_{2,2} & \cdots & p_{2,k} & 0 & 1 & \cdots & 0 \\ & & \vdots & & & & \vdots & \\ p_{n-k,1} & p_{n-k,2} & \cdots & p_{n-k,k} & 0 & 0 & \cdots & 1 \end{bmatrix}.$$

Since columns of $\mathbf{H}$ form an overcomplete basis, a given function can be represented in terms of this basis in many ways. Important is that the covering property of the code guarantees that the set of possible representations will necessarily include one with at most r columns. These representations can be determined by the following search procedure.

Let $\mathbf{s} = [s_1, \ldots, s_{n-k}]$ be an arbitrary vector that can be represented, and $\mathbf{v} = [v_1, \ldots, v_n]^T$ the vector of coefficients in this representation. Then $\mathbf{s} = \mathbf{Hv}$, where $w(\mathbf{v}) \leq r$. Now,

$$\mathbf{s} = \mathbf{Hv} = [\mathbf{P}^T | \mathbf{I}_{n-k}]\mathbf{v}$$

$$= \mathbf{P}^T \begin{bmatrix} v_1 \\ \vdots \\ v_k \end{bmatrix} + \begin{bmatrix} v_{k+1} \\ \vdots \\ v_n \end{bmatrix}.$$

Thus,

$$\begin{bmatrix} v_{k+1} \\ \vdots \\ v_n \end{bmatrix} = \mathbf{s} - \mathbf{P}^T \begin{bmatrix} v_1 \\ \vdots \\ v_k \end{bmatrix}. \tag{2.16}$$

Therefore, it is enough to check

$$1 + \binom{k}{1} + \binom{k}{2} + \cdots + \binom{k}{r}$$

vectors $\mathbf{v}' = [v_1, \ldots, v_k]^T$ with weight $1 \le w(\mathbf{v}) \le r$.

Example 38 *Consider the $(8, 4)$ binary code specified by the parity check matrix*

$$\mathbf{H} = \begin{bmatrix} 0 & 1 & 1 & 1 & 1 & 0 & 0 & 0 \\ 1 & 0 & 1 & 1 & 0 & 1 & 0 & 0 \\ 1 & 1 & 0 & 1 & 0 & 0 & 1 & 0 \\ 1 & 1 & 1 & 0 & 0 & 0 & 0 & 1 \end{bmatrix}.$$

It is easy to check that any vector $\mathbf{s}$ of length 4 can be expressed as a sum of at most 2 columns of $\mathbf{H}$ and, thus, the code has covering radius 2. To directly check all possibilities requires $1 + \binom{8}{1} + \binom{8}{2} = 37$ cases. Writing

$$\begin{aligned} v_5 &= v_2 + v_3 + v_4 + s_1, \\ v_6 &= v_1 + v_3 + v_4 + s_2, \\ v_7 &= v_1 + v_2 + v_4 + s_3, \\ v_8 &= v_1 + v_2 + v_3 + s_4, \end{aligned} \tag{2.17}$$

we see that $\binom{4}{1} + \binom{4}{2} = 10$ cases suffice.

The above problem is closely related to (syndrome based) decoding linear codes. This example suggests the following algorithm to determine the functional expression for a given function f from the selected code. Denote by $\mathbf{h}_i, i \in \{1, \ldots, n\}$ columns of the $((n - k) \times n)$ matrix $\mathbf{H}$. We want to represent a given function f with the function vector of length $n - k$ in terms of at most r columns of $\mathbf{H}$. Thus,

$$f = \sum_{i=0}^{r} \mathbf{h}_{e(i)},$$

where an empty sum denotes the zero-vector and, in general, we need to determine $e(i)$.

Algorithm 1 *(to find a minimal set of columns)*
Begin
If $f = 0$, zero columns needed. Stop.
else
$$w = r, l = 1$$

> *while $l \leq w$*
> > *generate vectors $[v_1, \ldots, v_k]$ of weight l*
> > *for each compute $[v_{k+1}, \ldots, v_n]$ from (2.17)*
> > > *if $\sum v_i < w$, then $w \leftarrow \sum v_i$*
> >
> > *$l \leftarrow l + 1$*
>
> *end while*

output $[v_1, \ldots v_n]$ where the non-zero give h_i.

End.

The application of this algorithm will be illustrated by the following examples.

Example 39 *Consider the parity check matrix of Example 38 and assume that the function vector to be represented is $[1, 0, 0, 1]$. Letting $[v_1, v_2, v_3, v_4]$ run through vectors of weight 1 and computing $[v_5, v_6, v_7, v_8]$, we get*

$$[v_1, v_2, v_3, v_4] = [1, 0, 0, 0], [0, 1, 0, 0], [0, 0, 1, 0], [0, 0, 0, 1]$$
$$[v_5, v_6, v_7, v_8] = [1, 1, 1, 0], [0, 0, 1, 0], [0, 1, 0, 0], [0, 1, 1, 1]$$

which gives a minimal representation as $h_2 + h_7$. Obviously another minimal representation is $h_3 + h_6$.

Example 40 *Consider the $(10, 5)$-code with the covering radius 2 whose parity check matrix is*

$$
\mathbf{H} = \begin{bmatrix}
1 & 1 & 0 & 0 & 0 & 1 & 0 & 0 & 0 & 0 \\
1 & 0 & 1 & 0 & 0 & 0 & 1 & 0 & 0 & 0 \\
1 & 0 & 0 & 1 & 0 & 0 & 0 & 1 & 0 & 0 \\
1 & 0 & 0 & 0 & 1 & 0 & 0 & 0 & 1 & 0 \\
0 & 1 & 1 & 1 & 1 & 0 & 0 & 0 & 0 & 1
\end{bmatrix}.
$$

Thus, each binary function $f(x)$ of a five valued variable $x \in \{0, 1, 2, 3, 4\}$ can be represented as

$$f = h_i + h_j,$$

where h_i and h_j are appropriately selected columns of $\mathbf{H}$. It is clear that out of $2^5 = 32$ functions there are 11 functions which can be represented by zero or one column of $\mathbf{H}_{(10,5)}$. For many functions we may have multiple solutions with two columns. The representations are summarized in Table 2.21 where $i + j$ denotes the sum of columns $h_i + h_j$.

Table 2.21: Representation of $f(x), x \in \{0, 1, 2, 3, 4\}$ in terms of columns of **H**

f	$h_i + h_j$	f	$h_i + h_j$
0.	-	16.	6, 2+10
1.	10, 2+6, 3+7, 4+8, 5+9	17.	2, 6+10
2.	9, 5+10	18.	2+5, 6+9
3.	5, 9+10	19.	2+9, 5+6
4.	8, 4+10	20.	2+4, 6+8
5.	4, 8+10	21.	2+8, 4+6
6.	4+5, 8+9	22.	1+7
7.	4+9, 5+8	23.	1+3
8.	7, 3+10	24.	2+3, 6+7
9.	3, 7+10	25.	2+7, 3+6
10.	3+5, 7+9	26.	1+8
11.	3+9, 5+7	27.	1+4
12.	3+4, 7+8	28.	1+9
13.	3+8, 4+7	29.	1+5
14.	1+6	30.	1
15.	1+2	31.	1 + 10

Example 41 *Consider the (23, 12) Golay code which has the covering radius 3. The parity check matrix is* $\mathbf{H} = [\mathbf{P}|\mathbf{I}_{11}]$ *where* $\mathbf{P}$ *is a* (11×12) *matrix defined as*

$$
\mathbf{P} = \begin{bmatrix}
1 & 0 & 0 & 1 & 1 & 1 & 0 & 0 & 0 & 1 & 1 & 1 \\
1 & 0 & 1 & 0 & 1 & 1 & 0 & 1 & 1 & 0 & 0 & 1 \\
1 & 0 & 1 & 1 & 0 & 1 & 1 & 0 & 1 & 0 & 1 & 0 \\
1 & 0 & 1 & 1 & 1 & 0 & 1 & 1 & 0 & 1 & 0 & 0 \\
1 & 1 & 0 & 0 & 1 & 1 & 1 & 0 & 1 & 1 & 0 & 0 \\
1 & 1 & 0 & 1 & 0 & 1 & 1 & 1 & 0 & 0 & 0 & 1 \\
1 & 1 & 0 & 1 & 1 & 0 & 0 & 1 & 1 & 0 & 1 & 0 \\
1 & 1 & 1 & 0 & 0 & 1 & 0 & 1 & 0 & 1 & 1 & 0 \\
1 & 1 & 1 & 0 & 1 & 0 & 1 & 0 & 0 & 0 & 1 & 1 \\
1 & 1 & 1 & 1 & 0 & 0 & 0 & 0 & 1 & 1 & 0 & 1 \\
0 & 1 & 1 & 1 & 1 & 1 & 1 & 1 & 1 & 1 & 1 & 1
\end{bmatrix}.
$$

Since the covering radius of the code is 3, the code can be used to represent binary functions of length 11 or, as will be explained later, it is a module to represent larger binary functions of lengths multiples of 11.

There are $2^{11} = 2048$ functions. The number of combinations of no more than three columns of the matrix $\mathbf{H}_{(11,23)}$ is

$$1 + \binom{23}{1} + \binom{23}{2} + \binom{23}{3} = 2048.$$

Thus, we have a unique representations for each function (which comes from the feature that this is a perfect code), where 23 functions are represented by a single column, 253 with two columns, while 1771 functions require three columns. For instance, the function $\mathbf{F} = [0, 0, 1, 1, 1, 0, 1, 0, 1, 1, 1]^T$ is represented as the sum of columns h_{10}, h_{11}, and h_{23}.

If the matrix $\mathbf{H}$ is replaced by the matrix

$$\mathbf{H}' = \begin{bmatrix} 1 & 0 \ldots 0 \\ \hline 0 & \\ \vdots & \mathbf{H} \\ 0 & \end{bmatrix}$$

we get a code that can be used to derive functional expressions for functions defined in 12 points, which can be viewed as functions of two binary and one ternary variable ($f(x_1, x_2, x_3)$, $x_1, x_2 \in \{0, 1\}$, $x_3 \in \{0, 1, 2\}$).

2.9.3 TERNARY GOLAY CODE

In this section, we apply the above to represent ternary functions.

Consider the code C over $GF(3)$ with the generator matrix

$$\mathbf{G} = \begin{bmatrix} 1 & 0 & 0 & 0 & 0 & 0 & 1 & 1 & 1 & 1 & 1 \\ 0 & 1 & 0 & 0 & 0 & 0 & 0 & 1 & -1 & -1 & 1 \\ 0 & 0 & 1 & 0 & 0 & 0 & 1 & 0 & 1 & -1 & -1 \\ 0 & 0 & 0 & 1 & 0 & 0 & -1 & 1 & 0 & 1 & -1 \\ 0 & 0 & 0 & 0 & 1 & 0 & -1 & -1 & 1 & 0 & 1 \\ 0 & 0 & 0 & 0 & 0 & 1 & 1 & -1 & -1 & 1 & 0 \end{bmatrix}$$

$$= \begin{bmatrix} 1 & 0 & 0 & 0 & 0 & 0 & 1 & 1 & 1 & 1 & 1 \\ 0 & 1 & 0 & 0 & 0 & 0 & 0 & 1 & 2 & 2 & 1 \\ 0 & 0 & 1 & 0 & 0 & 0 & 1 & 0 & 1 & 2 & 2 \\ 0 & 0 & 0 & 1 & 0 & 0 & 2 & 1 & 0 & 1 & 2 \\ 0 & 0 & 0 & 0 & 1 & 0 & 2 & 2 & 1 & 0 & 1 \\ 0 & 0 & 0 & 0 & 0 & 1 & 1 & 2 & 2 & 1 & 0 \end{bmatrix}.$$

C is the ternary Golay code. It is a perfect two-error correcting code and so also has covering radius $q = 2$.

We can write its parity check matrix as

$$
\mathbf{H} = \left[\begin{array}{rrrrrr|rrrrr}
-1 & 0 & -1 & 1 & 1 & -1 & 1 & 0 & 0 & 0 & 0 \\
-1 & -1 & 0 & -1 & 1 & 1 & 0 & 1 & 0 & 0 & 0 \\
-1 & 1 & -1 & 0 & -1 & 1 & 0 & 0 & 1 & 0 & 0 \\
-1 & 1 & 1 & -1 & 0 & -1 & 0 & 0 & 0 & 1 & 0 \\
-1 & -1 & 1 & 1 & -1 & 0 & 0 & 0 & 0 & 0 & 1
\end{array}\right]
$$

$$
= \left[\begin{array}{rrrrrr|rrrrr}
2 & 0 & 2 & 1 & 1 & 2 & 1 & 0 & 0 & 0 & 0 \\
2 & 2 & 0 & 2 & 1 & 1 & 0 & 1 & 0 & 0 & 0 \\
2 & 1 & 2 & 0 & 2 & 1 & 0 & 0 & 1 & 0 & 0 \\
2 & 1 & 1 & 2 & 0 & 2 & 0 & 0 & 0 & 1 & 0 \\
2 & 2 & 1 & 1 & 2 & 0 & 0 & 0 & 0 & 0 & 1
\end{array}\right]
$$

$$
= [-\mathbf{P}^T | \mathbf{I}_5].
$$

Let $\mathbf{s} = [s_1, s_2, s_3, s_4, s_5]^T$ be a ternary vector. Now, $\mathbf{s}$ can be represented as a linear combination of at most 2 columns of $\mathbf{H}$. Let again $\mathbf{v} = [v_1, v_2, \ldots, v_6, v_7, \ldots, v_{11}]^T$ be the vector indicating the coefficients of such a representation. Write $\mathbf{v}_1 = [v_1, v_2, \ldots, v_6]^T$ and $\mathbf{v}_2 = [v_7, v_8, \ldots, v_{11}]^T$. We have

$$
\mathbf{s} = \mathbf{H}\mathbf{v} = -\mathbf{P}^T \mathbf{v}_1 + \mathbf{v}_2.
$$

Considering this as a system of linear equations for $v_1, v_2, \ldots, v_{11}$, we see that solutions are obtained by letting v_i run through $GF(3^6)$. Because of the covering property of the code also the unique representation with at most two columns is among them. As those solutions with at least 3 nonzero v_i's are not of interests, it is enough to check $\binom{6}{1} \cdot 2 + \binom{6}{2} \cdot 2^2 = 72$ vectors $\mathbf{v}_1 = [v_1, \ldots, v_6]^T$ of weight at most 2. Direct check would require $1 + \binom{11}{1} \cdot 2 + \binom{11}{2} \cdot 2^2 = 243$ checks.

2.9.4   OCTACODE—A QUATERNARY CODE

We consider the octacode as an example of codes which can be used to derive sparse representations for quaternary functions. This code can be obtained by encoding pairs of binary values in the binary Nordstrom-Robinson code by quaternary values as integers modulo 4, that is in Z_4. It is a self-dual code, and can be defined by the generating matrix

$$
\mathbf{G} = [\mathbf{I}_4 | \mathbf{P}]
$$

$$
= \left[\begin{array}{rrrr|rrrr}
1 & 0 & 0 & 0 & 2 & 3 & 3 & 3 \\
0 & 1 & 0 & 0 & 1 & 2 & 3 & 1 \\
0 & 0 & 1 & 0 & 1 & 1 & 2 & 3 \\
0 & 0 & 0 & 1 & 1 & 3 & 1 & 2
\end{array}\right].
$$

This code has the covering radius 2, thus, each quaternary function of a single variable $f(x)$, $x \in \{0, 1, 2, 3\}$ can be represented as a sum of at most two columns of the matrix **G**.

There are $4^4 = 256$ functions to represent. For the quaternary case there are three non-zero weights for each column 1, 2, 3. Thus, there are

$$1 + 3 \cdot \binom{8}{1} + 9 \cdot \binom{8}{2} = 276$$

combinations. There are 25 functions which are multiplies of a single column of **G** and 232 functions which can be represented by the linear combination of two columns, 7 of which have different representations. These are functions specified by the function vectors [0022], [0202], [0220], [2020], [2002], [2200], and [2222].

2.9.5   SPARSE REPRESENTATIONS OBTAINED FROM ALGEBRAIC CODES

Besides the search algorithm presented above, it is possible to define analytical procedures which determine coefficients v_i, i.e., select the columns which represent the function. This will be illustrated by discussing sparse representations obtained from algebraic codes, in particular BCH-codes (Bose, Ray-Chaudhuri, Hocquenghem).

All the concepts that are discussed below can be formulated in much more general and also powerful ways. To fix ideas, we consider the simplest cases and present some concrete examples illustrating the approach.

Consider the finite field $GF(2^m)$ and let $n = 2^m - 1$. Then, n is the order of the cyclic group of units of $GF(2^m)$ and we know that there is a primitive element $\beta \in GF(2^m)$ such that

$$GF(2^m) = \{0, 1 = \beta^{n-1}, \beta, \beta^2, \ldots, \beta^{n-2}\}.$$

$GF(2^m)$ is a linear space of dimension m over $GF(2)$ and each element of $GF(2^m)$ can be identified with a binary column vector of length m. The primitive element is a root of an irreducible polynomial of degree m over $GF(2)$ and using this polynomial the multiplication table of the field $GF(2^m)$ is readily constructed.

Example 42 $x^3 + x + 1$ *is irreducible over* $GF(2)$ *and let* β *be its root. Because* $x^3 + x + 1$ *is irreducible over* $GF(2)$, *the elements* $1, \beta, \beta^2$ *are linearly independent over* $GF(2)$ *and as they span a space of dimension 3, i.e., of 8 elements it must be* $GF(2^3)$. *Writing*

$$1 = \begin{bmatrix} 1 \\ 0 \\ 0 \end{bmatrix}, \beta = \begin{bmatrix} 0 \\ 1 \\ 0 \end{bmatrix}, \beta^2 = \begin{bmatrix} 0 \\ 0 \\ 1 \end{bmatrix},$$

and using the fact that $\beta^3 = \beta + 1$, *we recursively find*

$$\beta^3 = \begin{bmatrix} 1 \\ 1 \\ 0 \end{bmatrix}, \beta^4 = \begin{bmatrix} 0 \\ 1 \\ 1 \end{bmatrix}, \beta^5 = \begin{bmatrix} 1 \\ 1 \\ 1 \end{bmatrix}, \beta^6 = \begin{bmatrix} 1 \\ 0 \\ 1 \end{bmatrix}.$$

Consider now the space of all binary polynomials $f(x)$ of degree at most 6 such that $f(\beta) = 0$. Thus,

$$f_0 + f_1\beta + f_2\beta^2 + \cdots + f_6\beta^6 = 0,$$

which is equivalent to

$$
\begin{bmatrix}
1 & 0 & 0 & 1 & 0 & 1 & 1 \\
0 & 1 & 0 & 1 & 1 & 1 & 0 \\
0 & 0 & 1 & 0 & 1 & 1 & 1
\end{bmatrix}
\begin{bmatrix}
f_0 \\ f_1 \\ f_2 \\ f_3 \\ f_4 \\ f_5 \\ f_6
\end{bmatrix}
=
\begin{bmatrix}
0 \\ 0 \\ 0
\end{bmatrix}.
$$

It is evident that this (code) space is equivalent to the Hamming code of length 7.

Let m and n be as above and $\delta = qt + 1 (\leq\leq n)$. A binary primitive BCH code with designated distance δ is the code C with the parity check matrix

$$
\mathbf{H} =
\begin{bmatrix}
1 & \beta & \beta^2 & \cdots & \beta^{n-1} \\
1 & \beta^2 & \beta^4 & \cdots & \beta^{(n-1)2} \\
& \vdots & & \vdots & \\
1 & \beta^{(\delta-1)} & \beta^{2(\delta-1)} & \cdots & \beta^{(n-1)(\delta-1)}
\end{bmatrix}.
\tag{2.18}
$$

Much is known about BCH-codes. For instance, for small values of n, δ, the covering radius is either known or there are good bounds on it. What is most interesting from our point of view is that there are efficient decoding algorithms. It means that the parity check matrices of BCH-codes provide overcomplete bases for Boolean functions with efficient algorithms to find a sparse representation.

Notice that in a field with the characteristic 2, we have

$$(a + b)^2 = a^2 + 2ab + b^2 = a^2 + b^2,$$

which means that $f(\beta^{2k}) = (f(\beta^k))^2$.

Thus, the parity check matrix (2.18) contains superfluous rows that should be removed before the matrix is used for sparse representations.

We outline the procedure for finding the coefficients of the sparse representation with an example.

Let β be the primitive element of $GF(2^m)$. We can form the parity check matrix of the BCH-code of length $15 = 2^4 - 1$ and designed distance 5 as

$$
\mathbf{H} =
\begin{bmatrix}
1 & \beta & \beta^2 & \cdots & \beta^{14} \\
1 & \beta^2 & \beta^4 & \cdots & \beta^{2\cdot14} \\
1 & \beta^3 & \beta^6 & \cdots & \beta^{3\cdot14} \\
1 & \beta^4 & \beta^8 & \cdots & \beta^{4\cdot14}
\end{bmatrix}.
$$

Now, the second and fourth rows are redundant and the parity check matrix is

$$\mathbf{H} = \begin{bmatrix} 1 & \beta & \beta^2 & \cdots & \beta^{14} \\ 1 & \beta^3 & \beta^6 & \cdots & \beta^{3\cdot14} \end{bmatrix},$$

where each of the two rows as elements of $GF(2^4)$ correspond to four binary rows. It is known that this code has covering radius 3.

Assume that $\mathbf{v}$ is an a arbitrary binary vector $\mathbf{v} = [\mathbf{v}_1^T, \mathbf{v}_2^T]^T$ of length 8. Then, $\mathbf{v}_1$ and $\mathbf{v}_2$ are binary column vectors of length 4 and so they correspond to unique elements of $GF(2^4)$, say s_1 and s_3. Representing $\mathbf{v}$ as a sum of at most three columns of $\mathbf{H}$ is equivalent to having a polynomial

$$e(x) = e_0 + e_1 x + \cdots + e_{14} x^{14},$$

with at most three non-zero terms such that

$$\begin{aligned} e(\beta) &= s_1, \\ e(\beta^3) &= s_3. \end{aligned}$$

Because we work over a field of characteristic 2, we also have

$$\begin{aligned} e(\beta^2) &= s_2 = s_1^2, \\ e(\beta^4) &= s_4 = s_1^4. \end{aligned}$$

It is evidently enough to be able to locate the non-zero coefficients of e. To this end, write

$$M = \{i \,|\, e_i \neq 0\}.$$

We know that $|M| \leq 3$. Denote

$$\sigma(z) = \prod_{i \in M} (1 + \beta^i z),$$

the so-called error locator and by

$$w(z) = \sum_{i \in M} \beta^i z \prod_{j \neq i} (1 + \beta^j z).$$

Now,

$$\begin{aligned} \frac{w(z)}{\sigma(z)} &= \sum_{i \in M} \frac{\beta^2 z}{1 + \beta^i z} = \sum_{i \in M} \left(\sum_{j=1}^{\infty} (\beta^i z)^j \right) \\ &= \sum_{j=1}^{\infty} z^j \underbrace{\sum_{i \in M} \beta^{ij}}_{e(\beta^j)} = \sum_{j=1}^{\infty} e(\beta^j) z^j, \end{aligned}$$

as formal power series.

Now, we know $e(\beta^j)$ for $j = 1, 2, 3, 4$ and thus have the equation

$$\frac{w(z)}{\sigma(z)} \equiv s(z) = s_1 z + s_2 z^2 + s_3 z^3 + s_4 z^4 \quad \text{mod } z^5.$$

Noticing that $w(z) = z\sigma'(z)$, we have

$$\sigma_1 z + 2\sigma_2 z^2 + 3\sigma_3 z^3 + \cdots = s(z)(\sigma_0 + \sigma_1 z + \cdots) \quad \text{mod } z^5,$$

which leads to equations

$$
\begin{aligned}
\sigma_1 &= s_1, \\
(0 =) \quad 2\sigma_2 &= s_1\sigma_1 + s_2, \\
(\sigma_3 =) \quad 3\sigma_3 &= s_1\sigma_2 + s_2\sigma_1 + s_3, \\
(0 =) \quad 4\sigma_4 &= s_1\sigma_3 + s_2\sigma_2 + s_3\sigma_1 + s_4,
\end{aligned}
$$

from which we get $\sigma(z)$ and eventually the set M indicating the required columns.

2.9.6   EXTENSIONS TO FUNCTIONS OF ARBITRARY LENGTH

In the previous presentation, we consider representation of functions with function vectors of the length equal to the dimension of certain codes. Now we explain how the same method can be directly extended to represent functions having function vectors of an arbitrary length.

Example 43 *Consider a matrix over $GF(2)$*

$$\mathbf{A} = \begin{bmatrix} 1 & 0 & 0 & 1 \\ 0 & 1 & 0 & 1 \\ 0 & 0 & 1 & 1 \end{bmatrix}.$$

We see that any binary function of a ternary variable $f(x)$, $x \in \{0, 1, 2\}$, can be expressed as a sum of at most two columns of $\mathbf{A}$.

Consider

$$\mathbf{I}_n \otimes \mathbf{A} = \begin{bmatrix} \mathbf{A} & \mathbf{0} & \cdots & \mathbf{0} \\ \mathbf{0} & \mathbf{A} & \cdots & \mathbf{0} \\ & & \ddots & \\ \mathbf{0} & \mathbf{0} & \cdots & \mathbf{A} \end{bmatrix}.$$

It is clear that any binary column of length $3n$ can be expressed as a sum of $2n$ columns of $\mathbf{I}_n \otimes \mathbf{A}$. So, using $4n$ columns or an overcomplete basis, we save $1/3$ of the coefficients.

Notice that, since each function vector can be written as the sum of n vectors having just three non-zero values, the number of checks to determine the function expression is equal ($n \times$ the number of checks for $\mathbf{A}$).

The method explained by the previous example can be generalized to functions of various length since we can select the parity check matrices of different codes such that the product of the number of their rows is equal to the length of the function given. Another important feature is that in general the number of checks to find coefficients to represent a function of a large length increases linearly as a function of the number of checks for the constituent parity check matrices used.

Example 44 *For instance, there are codes with length 32, dimension 16 and covering radius 6. Thus, there is a (16×32) binary matrix such that any vector of length 16 can be expressed as a sum of at most 6 columns of the matrix. This is a useful result, as we can see from Table 2.22 that shows the number of four-variable switching functions requiring six and more terms in different functional expressions. For definition of these expressions we refer to [166].*

The argument above indicates that for functions of n variables; we can take an overcomplete basis with 2^{n+1} vectors and any vector of length 2^n can be expressed as a sum of at most $6 \cdot 2^{n-4}$ vectors. Also, from the structure of the matrix, we see that they are easy to generate.

Table 2.22: Number of four-variable switching functions requiring 6 and more terms in SOP, PPRM, FPRM, GRM, ESOP, and SOP expressions				
PPRM	FPRM	GRM	ESOP	SOP
51451	34360	56	24	4242

Example 45 *As shown in Example 41, any vector of length 12 can be expressed as sum of at most 4 column vectors of parity check matrix for the extended Golay code $\mathbf{G}_{(12 \times 24)}$. The same construction as above gives a $(3 \cdot 2^n \times 3 \cdot 2^{n+1})$ matrix*

$$\mathbf{I}_{2^{n-2}} \otimes \mathbf{G}_{(12 \times 24)}.$$

Therefore, any vector of length $3 \cdot 2^{n+1}$ can be generated using 2^n of the $3 \cdot 2^{n+1}$ columns of this matrix.

2.9.7   VERY LARGE FUNCTIONS—THE ASYMPTOTIC CASE

There is a simple expression for the asymptotic form of the best possible covering radius of codes with fixed ratio of dimension and length. This implies directly that we also know asymptotically the parameters of sparse representations of the above type.

Assume that n is the minimal number of vectors in the overcomplete basis to represent any truth vector of length m with at most αn columns. Then, asymptotically

$$n \sim \frac{m}{H_2(\alpha)},$$

where

$$H_2(\alpha) = -\alpha \log_2 \alpha - (1 - \alpha) \log_2(1 - \alpha).$$

This can be interpreted, for instance, so that to represent any vector of length m with one column we need all the 2^m columns to choose from. However, as soon as the number of columns in the representation increases (relatively to m) the number of vectors needed in the overcomplete basis drops drastically. This is illustrated in Fig. 2.1.

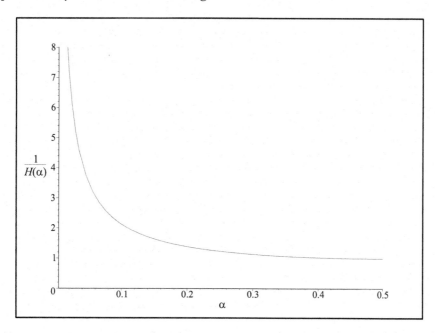

Figure 2.1: The asymptotic ration n/m as function of α, where n is the minimal size of the overcomplete basis for a space of dimension m to represent any vector as a linear combination with at most αn vectors.

CHAPTER 3

Spectral Representations of Multiple-Valued Functions

Spectral approach to analysis and synthesis of logic functions permits a uniform consideration of binary and multiple-valued logic functions by viewing them as elements of Hilbert spaces of functions on finite groups over the complex-field of some finite fields. Logic values are identified with integers or, more general as particular examples of complex numbers with imaginary parts equal to 0, or as elements of the corresponding finite fields. Spectral methods links Switching Theory and Digital Logic with Signal Processing and permit to use powerful theory of Abstract harmonic analysis and in particular Fourier theory on finite groups as well as some of related Signal Processing techniques to search for solutions of problems in these areas.

In general, the main advantage of spectral techniques is that they perform redistribution of the information contents of a signal modeled by a logic function and due to that some features of the signal hard to observe in the original domain become easily observable in the spectral domain. Another advantage is that spectral methods might facilitate some computationally demanding task and in this way make the related algorithms feasible in practice. Some other peculiar advantages can be obtained from spectral techniques depending on concrete tasks to be solved.

In this chapter, we will discuss spectral representations of logic functions on finite Abelian groups. For generalizations and extensions of methods to non-Abelian groups, we refer to [215].

3.1 FOURIER REPRESENTATIONS OF LOGIC FUNCTIONS

As noticed in Chapter 1, a logic function of n variables which do not necessarily take values in the same set, is defined as a mapping $f : \bigotimes_{i=1}^{n} S_i \to L$, where S_i are finite sets, $x_i \in S_i = \{0, 1, \ldots, p - 1\}$, $i = 1, 2, \ldots, n$, and L is a finite set assumed as the range for f, thus, $f \in L$. For each variable x_i, the domain S_i is usually enriched with an operation (often viewed as an addition) fulfilling the properties of a cyclic group G_i. In this way, the domain for f, $S = \bigotimes_{i=1}^{n} S_i$ acquires the structure of a group decomposable into the direct product of cyclic groups $G = \prod_{i=1}^{n} G_i$. It is usually assumed that $g_1 \leq g_2 \leq \ldots \leq g_n$, where $g_i = |G_i|$ is the order of G_i. For a group G of order g, we associate (permanently and bijectively) with each group element a non-negative integer from the set $\{0, 1, \ldots, g - 1\}$, and 0 is associated with the group identity. Thus, each group element will be identified with the fixed non-negative integer associated with it and with no other element.

The group operation $\circ$ of G can be expressed in terms of the group operations $\overset{\circ}{i}$ of the subgroups $G_i, i = 1, \ldots, n$ by

$$x \circ y = (x_1 \overset{\circ}{1} y_1, x_2 \overset{\circ}{2} y_2, \cdots, x_n \overset{\circ}{n} y_n), \quad x, y \in G, x_i, y_i \in G_i.$$

For the range L of f, the structure of a field P, that can be either the complex-field C or a finite field of order compatible with the cardinality of L is assumed. Therefore, a logic function $f(x_1, x_2, \ldots, x_n)$ is viewed as a function $f : G \to P$. The set of all logic functions for a given n expresses the structure of a linear (Hilbert) space $P(G)$ when enriched with the operation of addition borrowed from G and preformed componentwise and by the multiplication with a scalar in P.

Definition 22 *A mapping χ (not necessarily one-to-one) of a group G into the multiplicative group of (non-zero) complex numbers is called a homomorphism if, whenever $x_1 \circ x_2 = x_3$, then $\chi(x_1)\chi(x_2) = \chi(x_3), x_1, x_2, x_3 \in G$.*

Definition 23 *Any homomorphism of a group G into the multiplicative group of complex numbers is known as a character of G. The groups characters are functions $\chi_w : G \to C$ defined as*

$$\begin{aligned} |\chi_w(x)| &= 1, \\ \chi_w(x \circ y) &= \chi_w(x) \cdot \chi_w(y), \quad x, y \in G. \end{aligned}$$

If $\chi_w(x)$ is the character of G corresponding to the element w of G, then $\chi_0(x)$ is the principal character of G and it is defined as $\chi_0(x) = 1$ for any $x \in G$. It is a necessary element of the set of basis functions in terms of which the Fourier transforms on finite groups are defined, since it is required to capture the DC component in signals modeled by functions on finite groups.

To provide a method for construction of group characters of finite Abelian groups, we need the following notation and definitions.

Since any Abelian group G can be represented as a direct product of cyclic subgroups G_i, $i = 1, 2, \ldots, n$, as assumed above, it follows that there exist elements $\gamma_1, \gamma_2, \ldots, \gamma_n$ of the support set of G, $supp(G)$, such that for any $x \in supp(G)$ there exists $x_1, x_2, \ldots, x_n, 0 \leq x_i \leq N, i = 1, 2, \ldots, N, N$-the number of elements in $supp(G)$, such that

$$x = \gamma_1^{x_1} \circ \gamma_2^{x_2} \circ \cdots \circ \gamma_n^{x_n},$$

where $\circ$ is the group operation of G.

The elements γ_i are called *generators* and they form a basis of G. The dimension of the basis is n and generators are independent of x. It is assumed that for any x, the numbers $x_i, i = 1, \ldots, n$ are minimal. It is clear that if $g = |G|$ is the order of G, then $g_i = x_i + 1$ are orders of the constituent subgroups G_i of G.

Example 46 *Consider the group $G = (\{0, 1, 2, 3, 4, 5\}, \circ)$, where the group operation $\circ$ is defined in Table 3.1. The order of this group is $|G| = 6$, and the neutral (identity) element is $e = 0$. This group*

| **Table 3.1:** Group operation of G_6 in Example 46 |

G_6							G_1			G_2		

$\circ$	0	1	2	3	4	5
0	0	1	2	3	4	5
1	1	2	0	4	5	3
2	2	0	1	5	3	4
3	3	4	5	0	1	2
4	4	5	3	1	2	0
5	5	3	4	2	0	1

$\circ$	0	3
0	0	3
3	3	0

$\circ$	0	1	2
0	0	1	2
1	1	2	0
2	2	0	1

can be viewed as the direct product of the subgroups $G_1 = (\{0, 3\}, \circ)$, and $G_2 = (\{0, 1, 2\}, \circ)$, of orders $g_1 = |G_1| = 2$ and $g_2 = |G_2| = 3$, with generators $\gamma_1 = 3$ and $\gamma_2 = 1$, respectively.

Each element of G can be expressed in terms of generators γ_1 and γ_2 as

$$
\begin{aligned}
0 &= 3 \cdot 0 \circ 1 \cdot 0 \\
1 &= 3 \cdot 0 \circ 1 \cdot 1 \\
2 &= 3 \cdot 0 \circ 1 \cdot 2 \\
3 &= 3 \cdot 1 \circ 1 \cdot 0 \\
4 &= 3 \cdot 1 \circ 1 \cdot 1 \\
5 &= 3 \cdot 1 \circ 1 \cdot 2
\end{aligned}
$$

3.2   CONSTRUCTION OF GROUP CHARACTERS

Group characters form an orthogonal basis in terms of which the Fourier representations of functions on locally compact Abelian groups are defined [161]. Finite groups are compact; they are a subset of locally compact Abelian groups, thus, the same definition applies to functions defined on these groups. Therefore, in this section, we consider construction of group characters.

Definition 24 *Consider the Abelian group* $G = (\{0, 1, \ldots g - 1\}, \circ)$ *which is the direct product of* n *cyclic groups* G_i *of orders* $g_i, i = 1, 2, \ldots, n,$ *with generators* γ_i*. The* w*-th character of* G *is defined as*

$$
\chi_w(x) = exp \left(2\pi i \sum_{k=1}^{n} \frac{w_k x_k}{g_k} \right), \quad i = \sqrt{-1}, \tag{3.1}
$$

where

$$
\begin{aligned}
x &= \gamma_1^{x_1} \circ \gamma_2^{x_2} \circ \cdots \circ \gamma_n^{x_n}, \\
w &= \gamma_1^{w_1} \circ \gamma_2^{w_2} \circ \cdots \circ \gamma_n^{w_n},
\end{aligned}
$$

for $0 \leq w_i, x_i \leq g_i - 1$.

Table 3.2: Group characters for C_2, C_3, and C_4

C_2	C_3	C_4
$\chi_0 = [1, 1]$ $\chi_1 = [1, -1]$	$\chi_0 = [1, 1, 1]$ $\chi_1 = [1, e_1, e_2]$ $\chi_2 = [1, e_2, e_1]$	$\chi_0 = [1, 1, 1, 1]$ $\chi_1 = [1, i, -1, -i]$ $\chi_2 = [1, 1, -1, -1]$ $\chi_3 = [1, -i, -1, i]$

Table 3.3: Group characters for G_6

x, w	0	1	2	3	4	5
0	1	1	1	1	1	1
1	1	e_1	e_2	1	e_1	e_2
2	1	e_2	e_1	1	e_2	e_1
3	1	1	1	-1	-1	-1
4	1	e_1	e_2	-1	$-e_1$	$-e_2$
5	1	e_2	e_1	-1	$-e_2$	$-e_1$

Example 47 *Table 3.2 shows group characters for the cyclic groups $C_2 = (\{0, 1\}, \oplus_2)$, $C_3 = (\{0, 1, 2\}, \oplus_3)$, and $C_4 = (\{0, 1, 2, 3\}, \oplus_4)$, where C_4 is viewed as the group of integers modulo 4. Notice that the group characters of C_2 are the Walsh functions, while for C_3 and C_4 they are Vilenkin-Chrestenson functions, see for instance [99]. These groups are used as the domain groups for binary, ternary, and quaternary logic functions of a single variable. The direct products of each of these groups by itself n times produce the domain for functions of n variables. The combination of groups of different orders in the direct product is the domain for logic functions with variables of mixed cardinalities. For example, Table 3.3 shows the group characters for the group G_6 in Example 46. It is obvious that these characters are the Kronecker product of characters for C_2 and C_3. This group can be viewed as the domain group for functions of a binary variable $x_1 \in \{0, 1\}$ and a ternary variable $x_2 \in \{0, 1, 2\}$.*

The group characters, thus, the Vilenkin-Chrestenson functions also, under componentwise multiplication form a group isomorphic to the initial group G, whose characters they are. Thus, they are a complete orthogonal basis in the space of functions on G. This basis is used to define the Fourier transform on finite Abelian groups, since the classical Fourier transform is defined in terms of the exponential functions $exp(-jwx)$ which are the group characters of the real line R viewed as a locally compact Abelian group. Note that all finite groups are compact.

For the group $G = (\{0, 1, \dots, p - 1\}^n, \oplus_p)$, where for each subgroup G_i, the size $g_i = |G_i| = p$, $i = 1, 2, \dots, n$, the values w_i, x_i in (3.1) are components of the p-ary expansions of w, x, and the set of characters $\{\chi_w(x)\}$ is the system of Vilenkin-Chrestenson functions. In terms of this system, the Vilenkin-Chrestenson transform is defined for functions on G, i.e., functions of n variables taking values in the set $\{0, 1, \dots, p - 1\}$. Thus, it the case of such groups, we speak about the Vilenkin-Chrestenson transform and discuss it as a particular case of the Fourier transform

Table 3.4: Basic transform matrices for functions on C_2, C_3, and C_4

C_2	C_3	C_4
$\mathbf{W}(1) = \begin{bmatrix} 1 & 1 \\ 1 & -1 \end{bmatrix}$	$\mathbf{VC}_3(1) = \begin{bmatrix} 1 & 1 & 1 \\ 1 & e_2 & e_1 \\ 1 & e_1 & e_2 \end{bmatrix}$	$\mathbf{VC}_4(1) = \begin{bmatrix} 1 & 1 & 1 & 1 \\ 1 & -j & -1 & j \\ 1 & 1 & -1 & -1 \\ 1 & j & -1 & -j \end{bmatrix}$

on finite Abelian groups whose subgroups G_i are not necessarily identical as for instance for the group G_6 in Example 47. In the same context, the Vilenkin-Chrestenson transform is viewed as a generalization of the Walsh transform, that is the Fourier transform on C_2^n and used for functions of binary-valued variables, to functions on groups C_p^n, i.e., the functions of p-valued variables.

Definition 25 *For a function f defined on a finite Abelian group G, the Fourier transform is defined as*

$$S_f(w) = \sum_{i=1}^{n} f(x)\overline{\chi}_w(x),$$

where $\overline{\chi}_w(x)$ is the complex-conjugate transpose of $\chi_w(x)$.
The function f is reconstructed from its spectrum as

$$f(x) = \sum_{i=1}^{n} S_f(x)\chi_w(x).$$

In matrix notation, the basis functions (group characters) in terms of which the Fourier transform on a group G of order $g = |G|$ is defined, are represented as columns of a $(g \times g)$ matrix $[\chi_w(x)]$, $w, x = 0, 1, \ldots g - 1$. Then the corresponding transform matrices are the complex-conjugate transpose of $[\chi_w(x)]$.

Definition 26 *For a function f on a finite Abelian groups G of order $g = |G|$ represented by the function vector $\mathbf{F} = [f(0), \ldots, f(g-1)]^T$, the Fourier spectrum $\mathbf{S}_f = [S_f(0), \ldots, S_f(g-1)]^T$ is defined as*

$$\mathbf{S}_f = [\overline{\chi}_w(x)]\mathbf{F},$$

where $[\overline{\mathbf{Q}}]$ denotes the complex-conjugate transpose of a matrix $\mathbf{Q}$.

The direct product structure of the domain group G reflects into the Kronecker product structure of the set of characters. It follows that in this case, the Fourier transform matrix is defined as the Kronecker product of basic transform matrices, i.e., the Fourier transform matrices on the component subgroups G_i of G.

Example 48 *Table 3.4 shows the basic transform matrices for C_2, C_3, and C_4. It is obvious that these matrices correspond to the groups characters in Table 3.2 under complex-conjugation and transposition* *of matrices. Notice that in this case the matrices are symmetric, thus the transposition does not make a difference.*

The Walsh transform is defined in terms of the basic transform matrix on C_2. The Vilenkin-Chrestenson transform for ternary and quaternary logic functions is defined in terms of the transform matrices on C_3 and C_4. The Fourier transform on G_6 is defined by the transform matrix $Q = \mathbf{W}(1) \otimes \mathbf{VC}_3(1)$.

In some applications it is useful to encode function values $\{0, 1, \ldots, p - 1\}$ with the values of the character $\chi_1(x)$ to make the function to be processed similar to the basis functions in terms of which the transformation is defined. For instance, for $p = 2$, we do encoding $(0, 1) \rightarrow (1, -1)$ and for $p = 3$, the encoding $(0, 1, 2) \rightarrow (1, e_1, e_2)$. Such encoding results in peculiar properties of spectra of logic functions that can be useful in certain applications. For instance, this encoding ensures that we can get the maximum value coefficient p^n for the functions with highest correlation with rows of the transform matrices. Since these rows are isomorphic to the linear multiple-valued functions, we can estimate how far a given function is from the set of linear functions. Also, there are restrictions to the possible values of spectral coefficients of logic functions which facilitates controlling of correctness of the computation of spectral coefficients, etc. For more details on this subject, see [83], [100], [107], [114], [128], [129], [134], [130], [131], [132], [133], [135], [136], [137], [182].

Example 49 *Consider the ternary function $f(x_1, x_2) = x_1 x_2 \oplus 2x_1 \oplus 2x_2$. The function vector for f is $\mathbf{F} = [0, 2, 1, 2, 2, 2, 1, 2, 0]^T$, and the Vilenkin-Chrestenson spectrum is computed as*

$$
\begin{aligned}
3\mathbf{S}_f &= \mathbf{VC}_3(2)\mathbf{F} \\
&= \frac{1}{3}[12, 3e_2, 3e_1, 3e_2, 0, 3, 3e_1, 3, 0]^T.
\end{aligned}
$$

If we do the encoding $(0, 1, 2) \rightarrow (1, e_1, e_2)$, the function vector of f is $\mathbf{F} = [1, e_2, e_1, e_2, e_2, e_2, e_1, e_2, 1]^T$, and the Vilenkin-Chrestenson spectrum is

$$
\mathbf{S}_f = \frac{1}{3}[4, e_2, e_1, e_2, 0, 1, e_1, 1, 0]^T.
$$

Example 50 *Consider the quaternary function f whose GF-representation is $f(x_1, x_2) = x_1^2 x_2^2$. The function vector for f specified by the function vector $\mathbf{F} = [0, 0, 0, 0, 0, 1, 3, 2, 0, 3, 2, 1, 0, 2, 1, 3]^T$, and the Vilenkin-Chrestenson spectrum is computed as*

$$
\begin{aligned}
3\mathbf{S}_f &= \mathbf{VC}_4(2)\mathbf{F} \\
&= \frac{1}{16}[18, -6, -6, -6, -6, 2 + 4i, 2 - 4i, 2, -6, 2 - 4i, 2, 2 + 4i, \\
&\quad\; -6, 2 - 2 + 4i, 2 - 4i]^T.
\end{aligned}
$$

If we do the encoding $(0, 1, 2, 3) \rightarrow (1, i, -1, -i)$, *the function vector is* $\mathbf{F} = [1, 1, 1, 1, 1, i, -i, -1, 1, -i, -1, i, 1, -1, i, -i]^T$, *and the Vilenkin-Chrestenson spectrum is*

$$\mathbf{S}_f = \frac{1}{16}[4, 4, 4, 4, 4, 0, -4, 0, 4, -4, -4, 4, 4, 0, 4, -8]^T.$$

Example 51 *Consider the quaternary function* f *whose function vector is* $\mathbf{F} = [2, 0, 0, 1, 3, 3, 3, 2, 3, 3, 3, 0, 0, 1, 1, 2]^T$. *The Vilenkin-Chrestenson spectrum is computed as*

$$
\begin{aligned}
3\mathbf{S}_f &= \mathbf{VC}_4(2)\mathbf{F} \\
&= \frac{1}{16}[27, 1 - 2i, 3, 1 + 2i, -6 - 7i, 3i, -2 - 3i, 4 - 5i, \\
&\qquad -3, 3 - 2i, 5, 3 + 2i, -6 + 7i, 4 + 5i, -2 + 3i, -3i]^T.
\end{aligned}
$$

If we do the encoding $(0, 1, 2, 3) \rightarrow (1, i, -1, -i)$, *the function vector is* $\mathbf{F} = [-1, 1, 1, i, -i, -i, -i, -1, -i, -i, -i, 1, 1, i, i, -i]^T$, *and the Vilenkin-Chrestenson spectrum is*

$$
\begin{aligned}
\mathbf{S}_f &= \frac{1}{16}[1 - 3i, -3 - 3i, 1 - 3i, 1 + i, -5 + 5i, -1 + i, -1 + i, -1 + i, \\
&\qquad 3 - i, -5 + 3i, -5 + i, -1 - i, 5 + 3i, -3 - 5i, 1 - i, -3 + 3i]^T.
\end{aligned}
$$

3.3   HAAR SERIES FOR MULTIPLE-VALUED LOGIC FUNCTIONS

The subset of Vilenkin-Chrestenson functions consisting of functions whose index w is a multiple of a power of p, are the generalized Rademacher functions

$$K_{r,s}^{(p)}(x) = \chi_{r \cdot p^s}(x) = exp\left(\frac{2\pi i}{p} \cdot r \cdot s_s\right).$$

Example 52 *Table 3.5 shows the generalized Rademacher functions for* $p = 3$ *and* $n = 2$.

Table 3.5: Generalized Rademacher functions for $p = 3, m = 2$								
$K_{r,s}^{(3)}$	0	1	2	3	4	5	6	7	8
$K_{0,0}^{(3)}$	1	1	1	1	1	1	1	1	1
$K_{1,0}^{(3)}$	1	1	1	e_1	e_1	e_1	e_2	e_2	e_2
$K_{2,0}^{(3)}$	1	1	1	e_2	e_2	e_2	e_1	e_1	e_1
$K_{1,1}^{(3)}$	1	e_1	e_2	1	e_1	e_2	1	e_1	e_2
$K_{2,1}^{(3)}$	1	e_2	e_1	1	e_2	e_1	1	e_2	e_1

Definition 27 *The generalized Haar functions are defined in terms of the generalized Rademacher functions as*

$$M_{0,0}^{(p,1)}(x) \;=\; 1, for\ all\ x \in \{0, 1, \ldots, p^n - 1\} \tag{3.2}$$

$$M_{r,s}^{(p,q)}(x) \;=\; \begin{cases} K_{r,s}^{(p)}(x) & if\ x \in ((q - 1)p^{n+1-s}, q \cdot p^{n+1-s}) \\ 0, & otherwise. \end{cases} \tag{3.3}$$

CHAPTER 4

Decision Diagrams for Multiple-Valued Functions

Decision diagrams (DDs) are a specific data structure permitting compact representation, manipulation, and calculations with discrete functions taking relatively few different values. Decision diagrams are derived by the reduction of the corresponding decision trees.

From spectral interpretation [195], decision trees are graphical representations of the Fourier-like expansions of discrete functions with respect to some particular sets of basic functions. These basic functions are expressed through the labels at the edges. The values of constant nodes are the Fourier-like spectral coefficients determined with respect to the used basis. The explanation for this statement is the following.

A function expansion (decomposition rule) is performed at each non-terminal node of the decision tree. That expansion corresponds to a selected partial Fourier-like transform of f with respect to a variable in f. This is the variable used as the decision variable assigned to the considered node in the decision tree. In many decision trees, the same expansion is used for all the variables and all the nodes. In Kronecker decision trees, different expansions can be used for nodes at different levels in the decision tree. In a decision tree, a level consists of nodes at the same distance form the root node. The distance is measured as the length of the path from the root node to the considered node.

In Pseudo-Kronecker decision trees it is allowed to freely select the expansion rule for each node irrespectively to the expansion rules used in other nodes.

The outgoing edges of a node point to the subfunctions in the expansion of f performed in the node. The labels at the edges of a node are determined such that when the produced subfunctions are re-connected in terms of the labels (by the mapping inverse to that used in the decomposition) the function related to the incoming edge of the node is regained. Since we assign a given f to the decision diagram by preforming the selected functional expansions recursively with respect to all the variables, it follows that finally the constant nodes show the values of coefficients in the functional expansion for f. It follows that, conversely, a decision diagram shows a given function f in terms of the corresponding functional expansion or the Fourier-like expression in terms of the basis functions determined by labels at the edges along the paths from the root node to the constant nodes (the coefficients in the expansions).

The basis consisting of block pulses, i.e., represented by columns of the identity matrices of the corresponding orders, is included as a particular example. In another notation, this basis is represented by minterms of Boolean variables. It is used in the definition of the probably most widely known notion in the theory and practice of decision diagrams, the Binary decision diagrams (BDDs) [20] used to represent binary-valued logic functions. The same basis is used in Multi-terminal binary decision diagrams (MTBDDs) to represent integer or complex-valued functions of binary variables, and Multiple-place decision diagrams (MDDs) for the representation of multiple-valued logic functions. For definitions of these diagrams, see for instance [170] and the corresponding references therein.

This basis is somewhere called the trivial basis, since being represented by the identity matrix, it performs the identity mapping. It follows that the coefficients in the corresponding functional expression are function values. These coefficients, i.e., function values, are values of constant nodes in the decision diagrams.

Notice that in a complete disjunctive normal form, the products of variables (minterms) are disjoint, thus replacement of logic OR with logic EXOR is possible. In the reduced functional expressions, we work with implicants, i.e., products that do not necessarily contain all the variables, and such a replacement is impossible. Thus, BDD represents a function in the form of EXOR sum of products, where the products are obtained by multiplying labels at the edges along the paths from the root node to the constant nodes. The constant nodes in a BDD represent function values.

By selecting different sets of basis functions, various classes of decision diagrams other than BDDs are defined. For example, the Functional decision diagrams (FDDs) [101] are defined in terms of the Reed-Muller functions, that are the basis functions in the Reed-Muller transform. In FDDs, the values of constant coefficients are the Reed-Muller coefficients. Both these classes of decision diagrams, BDDs and FDDs, are examples of bit-level decision diagrams, since values of constant nodes are logic values 0 and 1. Many other classes of decision diagrams are defined in the same way with respect to various spectral transforms [195]. The spectral coefficients can be integers, real numbers, complex numbers, or matrices [140], [188], [191], [194], [212]. These are word-level diagrams since values of constant nodes are represented by computer words (bytes). As examples, we mention decision diagrams defined with respect to the Arithmetic and the Walsh transforms. In the Arithmetic spectral transform decision diagrams, the basis functions are of the same form as the Reed-Muller functions used in FDDs, however, with logic values 0 and 1 interpreted as integer 0 and 1. Walsh decision diagrams (WDDs) are defined in terms of the Walsh functions and the values of constant nodes are the Walsh spectral coefficients.

The reason for introducing a variety of these diagrams is that spectral transforms preserve the information content of a function to be represented, however, the distribution of information in the spectrum is different compared to that in the function vector. Thus, some regularities in the spectrum may appear, which are hardly observable in the initial function vector. These regularities, if they exist, result in a larger number of isomorphic subtrees in the decision tree defined with respect to the used spectral transform than in the decision tree defined with respect to the identity transform. Multiple copies of isomorphic subtrees can be viewed as the redundant information and, therefore, can be removed from the decision tree, resulting in the decision diagram, which is a reduced representation of the given function f.

4.1   DECISION TREES AND DECISION DIAGRAMS

Decision diagrams are obtained by reducing decision trees. The reduction of a decision tree is possible due to the existence of certain particular properties of functions to be represented and consists in removing the redundant information from the decision tree as will be briefly discussed bellow. Therefore, when discussing different classes of decision diagrams, i.e., diagrams defined in terms of different bases, alternatively, in terms of different expansion rules, it is better to consider first decision trees, since these considerations are independent of properties of particular functions.

As noticed above, the main idea behind defining decision diagrams in terms of various expansion rules is to increase diversity of different diagrams which increases the probability to find a diagram of small complexity to represent a given function f. The complexity is usually measured through the number of non-terminal nodes, although in some cases the number of constant nodes as well as complexity of interconnections are also taken into account. In this respect, the number of paths and the average path lengths are considered. Recall that a path in a decision diagram consists of edges that should be traversed to reach a constant node starting from the root node.

Construction of decision diagrams of smaller complexity for the given function is the same background idea that motivated the definition of Kronecker and Pseudo-Kronecker decision diagrams.

We will consider extensions and generalizations of the notion of decision diagrams to represent multiple-valued functions. The considerations will mainly relate to the diagrams representing the corresponding counterparts of BDD, Functional decision diagrams (FDDs), and Walsh decision diagrams for binary logic functions [170]. We will also consider the decision diagrams with attributed edges, by referring primarily to the Edge-valued BDDs [103] as the main representative of these large classes of diagrams in the binary base. Although we will try to provide all necessary definitions and explanations of the terminology and related notations, certain level of familiarity with these diagrams for binary functions can be useful in following the considerations presented.

4.2   REDUCTION RULES

Reduction rules used to derive a decision diagram from the decision tree are adapted to the decomposition rules performed at the nodes used in the tree [120]. In a general formulation, the possibility to delete or share a node relates to the existence of some isomorphic sub-trees in the decision tree. Isomorphic sub-trees correspond to the identical subfunctions in the function represented. Since a decision diagram is derived from the corresponding decision tree, the complexity of the decision diagram depends on the structure of the vector $\mathbf{V}$ representing the values of constant nodes in the decision tree. Under the structure of the vector $\mathbf{V}$ we consider the relationships among the subvectors $\mathbf{V}_k$ whose orders are determined by the group on which the represented discrete function is defined. In the case of the finite dyadic group $C_2^n = (\{0, 1\}^n, \oplus)$, where $\oplus$ denotes componentwise addition modulo 2, (logic EXOR), the orders of the subvectors $\mathbf{V}_k$ are 2^k, $k = 1, \ldots, n - 1$. For discrete functions on a finite Abelian group G, the orders of $\mathbf{V}_k$ are determined by the orders of the subgroups G_i to which G can be decomposed.

From spectral interpretation, entries of $\mathbf{V}$ are coefficients of Fourier or certain Fourier-like spectral transforms, and the complexity of a decision diagram depends on the Fourier-like spectrum of f, and in that way on the transform to which the corresponding decision tree relates. For a given f, different transforms provide decision diagrams of different complexity. Therefore, it is important to realize to which transforms various decision diagrams defined in literature are related. It is equally important to provide definitions of decision diagrams based upon other suitably chosen transforms. The efficiency of application of various already defined decision diagrams confirms such interest. For each defined decision tree and respectively decision diagram, there is always a class of functions for which that decision diagram is more suitable than any other decision diagram. Conversely, given a class of decision diagrams, it is always possible to find functions for which these diagrams have a large (even exponential) complexity.

4.3   MULTIPLE-PLACE DECISION DIAGRAMS

In binary logic, the Shannon expansion rule that is used in the definition of Binary decision diagrams (BDDs) is defined as

$$f(x_1, \ldots, x_n) = \overline{x}_i f_0(x_1, \ldots, x_{i-1}, 0, x_{i+1}, \ldots, x_n) \oplus x_i f_1(x_1, \ldots, x_{i-1}, 0, x_{i+1}, \ldots, x_n),$$

where $x_i \in \{0, 1\}$, and f_0 and f_1 are the co-factors of f with respect to x_i. A non-terminal node in a BDD is a graphic representation of the Shannon decomposition rule with the decision variable x_i and having outgoing edges pointing to f_0 and f_1.

The characteristic functions, defined in Chapter 2, Section 2.3, can be viewed as a generalization of the notion of literals to multiple-variables, and can be used to define a generalized Shannon expansion rule for multiple-valued functions. The following example illustrates the derivation of this notion for $p = 3$ by using the characteristic functions for ternary functions which are illustrated for $n = 2$ in Table 2.1. The generalizations to other values of p is straightforward.

Definition 28 *(Generalized Shannon expansion)*
The generalized Shannon expansion for ternary logic functions is defined as

$$f = J_0(x_i)f_0 + J_1(x_i)f_1 + J_2(x_i)f_2, \tag{4.1}$$

where f_i, $i = 0, 1, 2$ are the co-factors of f for $x_i \in \{0, 1, 2\}$.

The recursive application of the generalized Shannon expansion rule to all the variables in a given function f results in a functional expression which when graphically represented yields decision diagrams that are analog of the Binary decision diagrams (BDDs) [170] and represent a generalization of this concept to the representation of multiple-valued functions. These diagrams are called Multiple-place decision trees (MTDTs) and Multiple-place decision diagrams (MDDs) [177].

Example 53 *For $p = 3$ and $n = 2$, by expanding a given $f(x_1, x_2)$ with respect to x_1,*

$$f(x_1, x_2) = J_0(x_1)f(x_1 = 0, x_2) + J_1(x_1)f(x_1 = 1, x_2) + J_2(x_2)f(x_1 = 2, x_2).$$

After application of the generalized Shannon expansion with respect to x_2, it follows

$$
\begin{aligned}
f(x_1, x_2) &= J_0(x_2)(J_0(x_1)f(x_1 = 0, x_2 = 0) + J_1(x_1)f(x_1 = 1, x_2 = 0) \\
&\quad + J_2(x_1)f(x_1 = 2, x_2 = 0)) + J_1(x_2)(J_0(x_1)f(x_1 = 0, x_2 = 1) \\
&\quad + J_1(x_1)f(x_0 = 1, x_1 = 1) + J_2(x_0)f(x_0 = 2, x_1 = 1)) \\
&\quad + J_2(x_2)(J_0(x_1)f(x_1 = 0, x_2 = 2) + J_1(x_1)f(x_1 = 1, x_2 = 2) \\
&\quad + J_2(x_1)f(x_1 = 2, x_2 = 2) \\
&= J_0(x_2)J_0(x_1)f(x_1 = 0, x_2 = 0) + J_0(x_2)J_1(x_1)f(x_1 = 1, x_2 = 0) \\
&\quad + J_0(x_2)J_2(x_1)f(x_1 = 2, x_2 = 0) + J_1(x_2)J_0(x_1)f(x_1 = 0, x_2 = 1) \\
&\quad + J_1(x_2)J_1(x_1)f(x_1 = 1, x_2 = 1) + J_1(x_2)J_2(x_1)f(x_1 = 2, x_2 = 1) \\
&\quad + J_2(x_2)J_0(x_1)f(x_1 = 0, x_2 = 2) + J_2(x_2)J_1(x_1)f(x_1 = 1, x_2 = 2) \\
&\quad + J_2(x_2)J_2(x_1)f(x_1 = 2, x_2 = 2).
\end{aligned}
$$

Figure 4.1 shows the Multiple-place decision tree (MDT) for ternary functions of two variables.

For a function of n variables, the decision tree has $(n + 1)$ levels, each level consists of nodes to which the same variable is assigned. The first level has a single node called the root node. The level $(n + 1)$ consists of constant nodes showing the coefficients in the functional expressions whose graphical representation the decision trees are.

4.4 REDUCTION OF DECISION TREES

Decision diagrams are derived by the reduction of decision trees by eliminating the redundant information expressed in terms of the isomorphic subtrees, correspondingly subdiagrams. Reduction is accomplished by sharing isomorphic subtrees and deleting any redundant information in the decision tree. It is assumed that two subtrees are isomorphic if:

1. they are rooted in nodes at the same level; and

2. the constant nodes of the subtrees represent identical subvectors V_i in the vector of values of constant nodes V.

This definition includes different reduction rules used in different decision diagrams for either binary or the multiple-valued functions as well as bit-level and word-level decision diagrams [191]. In this connection, see also [89], [90], [118], [229].

The minimum possible isomorphic subtrees are equal constant nodes. In this case, the function represented has equal values at the points corresponding to these equal-valued constant nodes.

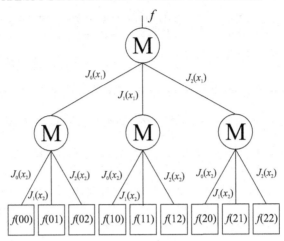

Figure 4.1: MDT for f in Example 53.

The maximum possible isomorphic subtrees are p equal subtrees rooted at the nodes pointed by the outgoing edges of the root node. In that case, the function f is independent on the variable assigned to the root node.

Definition 29 *(MDD reduction rules)*

1. *If descendent nodes of a node are identical, then delete the node and connect the incoming edges of the deleted node to the corresponding successor. The label of this incoming edge is re-determined as the product of the label at the initial incoming edge with the sum of labels at the outgoing edges of the deleted node.*

2. *Share isomorphic subtrees, i.e., if there are isomorphic subtrees, keep a single subtree and redirect to it the incoming edges of all other isomorphic subtrees.*

When a node is shared or deleted from the decision diagram, relabeling of the corresponding edges is required. In the general case, the relabeling is performed so that labels at the outgoing edges of the deleted node are added and their sum is multiplied by the sum of labels at the incoming edges to the deleted node. The summation and multiplication are preformed in the underlying algebraic structures used in the definition of the decision diagram. This way of relabeling the edges in a decision diagram is illustrated by the following examples illustrating various decision diagrams.

Definition 30 *(Cross points)*
Cross point is a point where an edge longer than one crosses a level in the decision diagram.

The cross points are useful to express the impact of deleted nodes, which is important to take into account in computing over decision diagrams or performing the realizations of functions represented by decision diagrams. A cross point is illustrated by the decision diagram in Fig. 4.3 for the function f in Example 54.

In a decision tree, edges connect nodes at successive levels, and we say that the length of such edges is 1. Due to the reduction, in a decision diagram, edges longer than one, i.e., connecting nodes at non-successive levels can appear. For example, the length of an edge connecting a node at the $(i-1)$-th level with a node at the $(i+1)$-th level is two.

Nodes to which the same decision variable is assigned form a level in the decision tree or the diagram.

A path consists of nodes at different levels, with a single node at each level, from the root node to a constant node. Thus, each path connects the root node, and a single node from each level including the level of constant nodes, i.e., a path consists of edges connecting a single node per level, and the length of the path is the sum of lengths of edges of which the path consists.

Example 54 *Figures 4.2 and 4.3 show the MDT and MDD for the function* $f = \sqrt{x_1 x_2}$ *over* $GF(4)$ *that is used as the Example 4 in [145]. The function vector* **F** *of* f *is* **F** = $[0, 0, 0, 0, 0, 1, 3, 2, 0, 3, 2, 1, 0, 2, 1, 3]^T$. *The cross point is shown to indicate the reduced node in the corresponding MDT.*

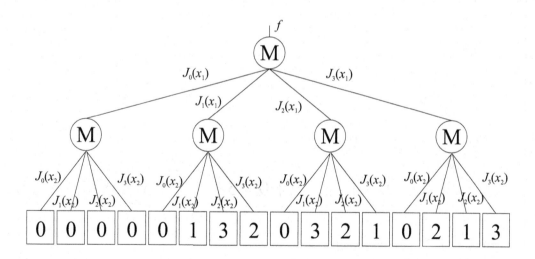

Figure 4.2: MDT for f in Example 54.

Example 55 *Figures 4.4 and 4.5 show the MDT and MDD for the two-variable function* f *over* $GF(4)$ *specified by the function vector* **F** = $[2, 1, 3, 0, 1, 3, 2, 0, 3, 2, 1, 0, 0, 0, 0, 0]^T$.

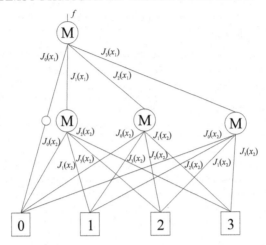

Figure 4.3: MDD for f in Example 54.

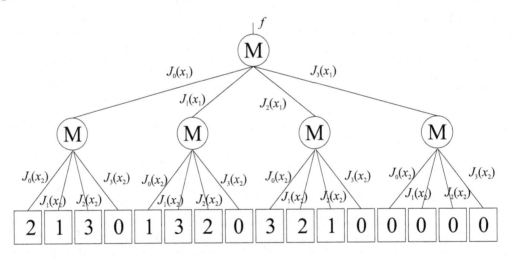

Figure 4.4: MDT for f in Example 55.

4.5 FUNCTIONAL DECISION DIAGRAMS FOR MV FUNCTIONS

As in the case of BDDs for binary switching functions, there are multiple-valued functions for which MDDs have a large complexity in the number of nodes or the number of edges. For this reason, functional decision diagrams for multiple-valued functions are defined as a generalization of Functional decision diagrams (FDDs) for binary functions [186], [189]. It follows that these

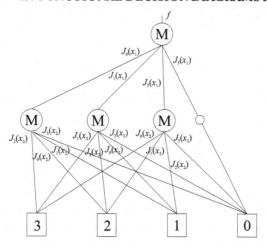

Figure 4.5: MDD for f in Example 55.

diagrams are defined in terms of functional expressions for multiple-valued functions representing generalizations of Reed-Muller expressions for binary functions. For instance, we will briefly present functional decision diagrams that are defined with respect to the decomposition rules defined in terms of the basic Galois field transform matrices and Reed-Muller-Fourier transform matrices for $p = 4$. A generalization to other values of p is straightforward.

4.5.1   GALOIS FIELD FUNCTIONAL DECISION DIAGRAMS

The multiple-valued equivalent of the positive Davio rule is defined from the Galois field transform in $GF(4)$ as follows.

The GF-expressions in $GF(4)$ are defined in terms of basis functions x^i, where $x, i \in \{0, 1, 2, 3\}$. If these functions $x^0 = 1$, x^1, x^2, and x^3 are written as columns of a (4×4) matrix over $GF(4)$, we get

$$\mathbf{X}_{4GF}(1) = \begin{bmatrix} 1 & 0 & 0 & 0 \\ 1 & 1 & 1 & 1 \\ 1 & 2 & 3 & 1 \\ 1 & 3 & 2 & 1 \end{bmatrix}.$$

The matrix inverse to $\mathbf{G}_{4GF}$ is the GF-transform matrix

$$\mathbf{G}_{4GF}(1) = \begin{bmatrix} 1 & 0 & 0 & 0 \\ 0 & 1 & 3 & 2 \\ 0 & 1 & 2 & 3 \\ 1 & 1 & 1 & 1 \end{bmatrix}.$$

Therefore, each quaternary function of a single variable can be viewed as a function over $GF(4)$ and represented as

$$f = \begin{bmatrix} 1 & x & x^2 & x^3 \end{bmatrix} \mathbf{G}_{4GF}(1)\mathbf{F},$$

where $\mathbf{F} = [f_0, f_1, f_2, f_3]^T$. From there, we get the positive Davio expansion for quaternary functions as

$$
\begin{aligned}
f &= f_0 \oplus x_i(f_1 \oplus 3f_2 \oplus 2f_3) \oplus x_i^2(f_1 \oplus 2f_2 \oplus 3f_3) \\
&\oplus x_i^3(f_0 \oplus f_1 \oplus f_2 \oplus f_3).
\end{aligned}
\tag{4.2}
$$

Definition 31 *The Galois field decision diagram is a decision diagram derived by the recursive application of the positive Davio GF-expression (4.2) recursively to all the variables in a function f to be represented. Each node in a GFDD realizes the mapping (4.2) and the values of constant nodes are the GF-coefficients of f. The outgoing edges of a node in the GFDD are labeled by $x_i^0 = 1$, x_i^1, x_i^2, and x_i^3.*

From this definition, a GFDD is the graphic representation of the GF-expression for f, and each path from the root node to a constant node corresponds to a product term in the GF-expression for f. The constant node to which the path terminates shows the inner product of f with the corresponding basis function in the GF-transform used.

Example 56 *[186] Figure 4.6 shows the GFDD for the function $f = \sqrt{x_1 x_2}$ over $GF(4)$ that is used as the Example 4 in [145]. The function vector $\mathbf{F}$ (Example 53) of f and the vector $\mathbf{S}_{f,4GF}$ of GF-coefficients of f are $\mathbf{S}_{f,4GF} = [0, 0, 0, 0, 0, 1, 3, 2, 0, 3, 2, 1, 0, 2, 1, 3]^T$ and $\mathbf{S}_{f,4GF} = [0, 0, 0, 0, 0, 0, 0, 0, 0, 0, 1, 0, 0, 0, 0, 0]^T$.*

4.5.2   KRONECKER GALOIS FIELD DECISION DIAGRAMS

The complements of a variable in $GF(p)$ are defined as $\overset{k-}{x}_i = x_i \oplus k$, $k = 0, 1, \ldots, p - 1$. Thus, for a single variable function in $GF(4)$ there are four polarities; the zero polarity and three negative polarities for $k = 1, 2, 3$.

1. The 1–4 negative Davio expansion in $GF(4)$ is defined as

$$
\begin{aligned}
f &= f_1 \oplus \overset{1-}{x}_i (f_0 \oplus 2f_2 \oplus 3f_3) \oplus \overset{1-2}{x}_i (f_0 \oplus 3f_2 \oplus 2f_3) \\
&\oplus \overset{1-3}{x}_i (f_0 \oplus f_1 \oplus f_2 \oplus f_3).
\end{aligned}
\tag{4.3}
$$

2. The 2–4 negative Davio expansion in $GF(4)$ is defined as

$$
\begin{aligned}
f &= f_2 \oplus \overset{2-}{x}_i (3f_0 \oplus 2f_1 \oplus f_3) \oplus \overset{2-2}{x}_i (2f_0 \oplus 3f_1 \oplus f_3) \\
&\oplus \overset{2-3}{x}_i (f_0 \oplus f_1 \oplus f_2 \oplus f_3).
\end{aligned}
\tag{4.4}
$$

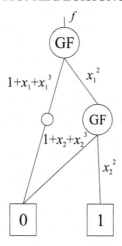

Figure 4.6: GFDD for f in Example 56.

3. The 3–4 negative Davio expansion in $GF(4)$ is defined as

$$f = f_3 \oplus \overset{2-}{x}_i \, (2f_0 \oplus 3f_1 \oplus f_2) \oplus \overset{2-2}{x}_i \, (3f_0 \oplus 2f_1 \oplus f_2)$$
$$\oplus \overset{2-3}{x}_i \, (f_0 \oplus f_1 \oplus f_2 \oplus f_3). \tag{4.5}$$

Definition 32 *A Kronecker Galois Field Decision Tree (KGFDT) is a decision tree associated to a given function by using the expansions (4.1), (4.2), (4.3), (4.4), (4.5), with the same expansion at a level in the decision tree.*

Definition 33 *Kronecker Galois Field Decision Diagrams (KGFDDs) are decision diagrams for representation of MV functions derived by the reduction of KGFDTs.*

KGFDDs derived by using the 4-positive Shannon expansion rule correspond to BDDs for switching functions. These derived by using 4-positive Davio expansion correspond to pFDDs (zero-polarity Reed-Muller DDs), while KGFDDs derived by using 4-positive Davio and $1 - 4$, $2 - 4$, $3 - 4$ negative Davio expansions correspond to FDDs (the Reed-Muller decision diagrams [170], [200]).

Example 57 *Figure 4.7 shows the KGFDD for the function f in Example 55, with the spectrum $\mathbf{S}_{f,4GF} = [2, 1, 3, 0, 1, 3, 2, 0, 3, 2, 1, 0, 0, 0, 0, 0]^T$, with $2 - 4$ nD nodes for x_1 and $3 - 4$ Davio nodes for x_2. This decision diagram represents f as*

$$f = \overset{3-}{x}_2 \oplus \overset{2-}{x}_1 \overset{3-}{x}_2 .$$

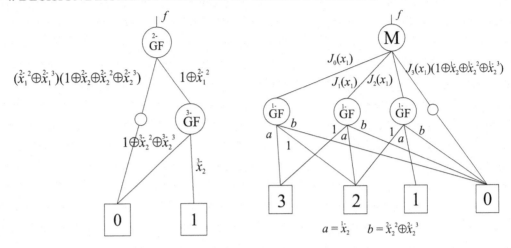

Figure 4.7: KGFDD for f in Example 57 with $2 - 4$ Davio node for x_1 and $3 - 4$ Davio node for x_2.

Figure 4.8: KGFDD for f in Example 58 with 4 Shannon node for x_1 ad $1 - 4$ Davio nodes for x_2.

Example 58 *Figure 4.8 shows a KGFDD for the function f from Example 55 with $\mathbf{F} = [2, 1, 3, 0, 1, 3, 2, 0, 3, 2, 1, 0, 0, 0, 0, 0]^T$ derived by using $4 - S$ nodes for x_1 and $1 - 4$ nD nodes for x_2. From this decision diagram, by traversing the paths and multiplying the labels at the edges the path consists of with the value of the constant node where the path terminates, we determine the functional expression for f corresponding to this decision diagram as*

$$ f = J_0(x_1)(1 \oplus 3 \overset{1-}{x}_2) \oplus J_1(x_1)(3 \oplus 2 \overset{1-}{x}_2) \oplus J_2(x_1)(2 \oplus \overset{1-}{x}_2). $$

4.6   REED-MULLER-FOURIER DECISION DIAGRAMS

The Reed-Muller-Fourier decision diagrams are defined in the same way as the GFDDs, however, by referring to the Reed-Muller-Fourier basic matrix to derive the Reed-Muller-Fourier decomposition rule [205]. For instance, any single variable function $f(x)$, x, $f \in \{0, 1, 2, 3\}$, can be represented as

$$ f = f_0 \oplus x(3f_0 \oplus f_1) \oplus x^{*2}(3f_0 \oplus 2f_1 \oplus 3f_2) \oplus x^{*3}(3f_0 \oplus 3f_1 \oplus f_2 \oplus f_3). \qquad (4.6) $$

Example 59 *The RMFDD of the function from Example 54 with $\mathbf{F} = [0, 0, 0, 0, 0, 1, 3, 2, 0, 3, 2, 1, 0, 2, 1, 3]^T$ is shown in Fig. 4.9. The vector of RMF-coefficients is given by $S_{f,4RMF} = [0, 0, 0, 0, 0, 1, 3, 0, 0, 3, 2, 0, 0, 0, 0, 2]^T$. Note that in this particular example, the number of nodes in the RMFDD is equal to that in the RMDD, however, the number of branches differs.*

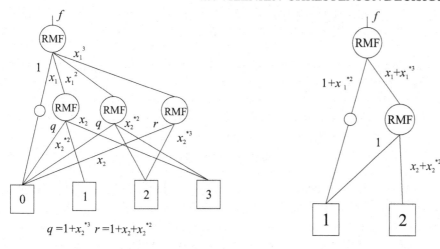

Figure 4.9: RMFDD for f in Example 59.

Figure 4.10: RMFDD for f in Example 60.

Example 60 *Figure 4.10 shows the RMFDD of the two–variable quaternary function f given by the function vector* $\mathbf{F} = [3, 0, 0, 0, 0, 3, 3, 3, 0, 2, 2, 2, 0, 0, 0, 0]^T$. *The vector of RMF-coefficients is given by* $\mathbf{S}_{f,4RMF} = [3, 3, 3, 3, 3, 2, 2, 2, 3, 3, 3, 3, 3, 2, 2, 2]^T$.

4.7   VILENKIN-CHRESTENSON DECISION DIAGRAMS

Spectral transform decision diagrams are defined in terms of decomposition rules derived from the basic transform matrices for various spectral transforms. A particular case are the Fourier decision diagrams defined in terms of Fourier transforms on finite groups [191], [212]. Such examples are Walsh decision diagrams (WDDs) for representation of functions in binary variables and Vilenkin-Chrestenson decision diagrams (VCDDs) that are their extension to functions in multiple-valued variables.

4.7.1   VILENKIN-CHRESTENSON DECISION DIAGRAMS FOR $p = 3$

The basic Vilenkin-Chrestenson transform matrix for $p = 3$ is defined as

$$\mathbf{VC}_3^{-1}(1) = \frac{1}{3} \begin{bmatrix} 1 & 1 & 1 \\ 1 & e_2 & e_1 \\ 1 & e_1 & e_2 \end{bmatrix}, \tag{4.7}$$

where $e_1 = exp(\frac{2\pi i}{3}) = -\frac{1}{2}(1 - i\sqrt{3})$, and $e_2 = \bar{e}_1 = exp(\frac{4\pi i}{3}) = -\frac{1}{2}(1 + i\sqrt{3})$, where $\bar{e}_1$ is the complex-conjugate of e_1.

Recall that given a matrix $\mathbf{R}$ specifying basis functions used to define a spectral transform, we use the inverse matrix $\mathbf{R}^{-1}$ to compute spectral coefficients. This explains the superindex -1 for $\mathbf{VC}^{-1}(1)$ in (4.7).

The basis functions $\chi(w, x)$, $w, x = 0, 1, 2$, for the Vilenkin-Chrestenson transform for $p = 3$ and $n = 1$ are defined as columns of the matrix

$$\mathbf{VC}_3(1) = \begin{bmatrix} 1 & 1 & 1 \\ 1 & e_1 & e_2 \\ 1 & e_2 & e_1 \end{bmatrix}. \tag{4.8}$$

Each single variable ternary function f specified by the function vector $\mathbf{F} = [f(0), f(1), f(2)]^T$ can be represented as

$$f = \frac{1}{3}(\chi(0, x)S_f(0) + \chi(1, x)S_f(1) + \chi(2, x)S_f(2)),$$

where, from (4.7),

$$\begin{array}{rcl} S_f(0) & = & f(0) + f(1) + f(2) \\ S_f(1) & = & f(0) + e_2 f(1) + e_1 f(2) \\ S_f(2) & = & f(0) + e_1 f(1) + e_2 f(2). \end{array}$$

The Vilenkin-Chrestenson functions for $p = 3$ can be expressed in terms of multiple-valued variables as

$$\begin{array}{rcl} \chi(0, x) & = & 1, \\ \chi(1, x) & = & 1 + a_1 x + a_2 x^2, \\ \chi(2, x) & = & 1 + \bar{a}_1 x + \bar{a}_2 x^2, \end{array}$$

where $\bar{a}$ is the complex-conjugate of a and $a_1 = -\frac{11}{4} + i\frac{5\sqrt{3}}{4}$, $a_2 = \frac{5}{4} - i\frac{3\sqrt{3}}{4}$, $x \in \{0, 1, 2\}$.

From this consideration we can derive the expansion rule whose recursive application define the Vilenkin-Chrestenson decision diagrams

$$f = \frac{1}{3}(1 \cdot S_f(0) + (1 + a_1 x + a_2 x^2)S_f(1) + (1 + \bar{a}_1 x + \bar{a}_2 x^2)S_f(2)). \tag{4.9}$$

Example 61 *Consider the two-variable function f specified by the function vector $\mathbf{F} = [0, 1, 2, 1, 2, 0, 2, 0, 1]^T$. The Vilenkin-Chrestenson spectrum is $S_f = \frac{1}{9}[9, 0, 0, 0, t, 0, 0, 0, \bar{t}]^T$, where $t = -4.5 + i2.6$, and $\bar{t} = -4.5 - i2.6$. If we do encoding $(0, 1, 2) \rightarrow (1, e_1, e_2)$, the function vector is $\mathbf{F}_e = [1, e_1, e_2, e_1, e_2, 1, e_2, 1, e_1]^T$. Then, the Vilenkin-Chrestenson spectrum is $S_f = [0, 0, 0, 0, 1, 0, 0, 0, 0]^T$. Fig. 4.11 shows a MDD for f and VCDDs for f before and after encoding.*

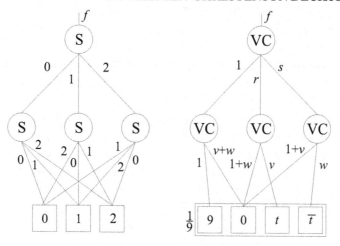

Figure 4.11: MDD and VCDD for f in Example 61.

In the VCDD for f, the values of constant nodes are $t = -4.5 + i2.6$ and $\bar{t} = -4.5 - i2.6$, while labels at the edges are

$$
\begin{array}{ll}
r = 1 + a_1 x_1 + a_2 x_1^2 & v = 1 + a_1 x_2 + a_2 x_2^2 \\
s = 1 + b_1 x_1 + b_2 x_1^2 & w = 1 + b_1 x_2 + b_2 x_2^2
\end{array}.
$$

The rectangle around the constant nodes means that all the values should be multiplied by $\frac{1}{9}$.

4.7.2   VILENKIN-CHRESTENSON DECISION DIAGRAMS FOR $p = 4$

The basic Vilenkin-Chrestenson transform matrix for $p = 4$ is defined as

$$
\mathbf{VC}_4^{-1}(1) = \frac{1}{4}
\begin{bmatrix}
1 & 1 & 1 & 1 \\
1 & -i & -1 & i \\
1 & -1 & 1 & -1 \\
1 & i & -1 & -i
\end{bmatrix}. \tag{4.10}
$$

This matrix serves as the kernel of the transform matrix for the Vilenkin-Chrestenson transform defined in terms of the basis functions that are generated by the Vilenkin-Chrestenson functions $\chi(w, x)$, $w, x \in \{0, 1, 2, 3\}$, which can be represented by columns of the matrix $\mathbf{VC}(1)$ inverse to $\mathbf{VC}^{-1}(1)$. Thus,

$$
\mathbf{VC}_4(1) =
\begin{bmatrix}
1 & 1 & 1 & 1 \\
1 & i & -1 & -i \\
1 & -1 & 1 & -1 \\
1 & -i & -1 & i
\end{bmatrix}. \tag{4.11}
$$

Each single-variable quaternary function f specified by the function vector $\mathbf{F} = [f(0), f(1), f(2), f(3)]^T$ can be represented as

$$f = \frac{1}{4}(\chi(0, x)S_f(0) + \chi(1, x)S_f(1) + \chi(2, x)S_f(2) + \chi(3, x)S_f(3)),$$

where, from (4.10),

$$\begin{array}{rcl}
S_f(0) & = & f(0) + f(1) + f(2) + f(3), \\
S_f(1) & = & f(0) - if(1) - f(2) + if(3), \\
S_f(2) & = & f(0) - f(1) + f(2) - f(3), \\
S_f(3) & = & f(0) + if(1) - f(2) - if(3).
\end{array}$$

The Vilenkin-Chrestenson functions can be expressed in terms of multiple-valued variables as

$$\begin{array}{rcl}
\chi(0, x) & = & 1, \\
\chi(1, x) & = & 1 + a_1 x + a_2 x^2 + a_3 x^3, \\
\chi(2, x) & = & 1 + b_1 x + b_2 x^2 + b_3 x^3, \\
\chi(3, x) & = & 1 + c_1 x + c_2 x^2 + c_3 x^3,
\end{array}$$

where

$$\begin{array}{lll}
a_1 = \frac{1}{3} - i, & a_2 = -\frac{2}{3}, & a_3 = -\frac{2}{3} + 2i, \\
b_1 = -\frac{2}{3}, & b_2 = \frac{4}{3}, & b_3 = -\frac{8}{3}, \\
c_1 = \frac{1}{3} + i, & c_2 = -\frac{2}{3}, & c_3 = -\frac{2}{3} - 2i.
\end{array}$$

Note that $c_i = \overline{a}_i, i = 0, 1, 2, 3$, where $\overline{a}$ is the complex-conjugate of a.

From this consideration we can derive the expansion rule whose recursive application define the Vilenkin-Chrestenson decision diagrams for $p = 4$

$$\begin{aligned}
f & = \frac{1}{4}(1 \cdot S_f(0) + (1 + a_1 x + a_2 x^2 + a_3 x^3)S_f(1) \\
& \quad + (1 + b_1 x + b_2 x^2 + b_3 x^3)S_f(2) + (1 + c_1 x + c_2 x^2 + c_3 x^3)S_f(3)). \quad (4.12)
\end{aligned}$$

Example 62 *Figure 4.12 shows the VCDD for the function $f = x_1 \oplus x_2$, where addition is the addition in $GF(4)$. Thus, this function is defined by the function vector*

$$\mathbf{F} = [0, 1, 2, 3, 1, 0, 3, 2, 2, 3, 0, 1, 3, 2, 1, 0]^T.$$

The Vilenkin-Chrestenson spectrum is

$$\mathbf{S}_f = [24, 0, 0, 0, 0, 8i, 0, -8, 0, 0, -8, 0, 0, -8, 0, -8i]^T.$$

Figure 4.12 shows the corresponding VCDD. In this figure,

$$\begin{array}{ll}
r = 1 + a_1 x_1 + a_2 x_1^2 + a_3 x_1^3, & v = 1 + a_1 x_2 + a_2 x_2^2 + a_3 x_2^3, \\
s = 1 + b_1 x_1 + b_2 x_1^2 + b_3 x_1^3, & w = 1 + b_1 x_2 + b_2 x_2^2 + b_3 x_2^3, \\
t = 1 + c_1 x_1 + c_2 x_1^2 + c_3 x_1^3, & y = 1 + c_1 x_2 + c_2 x_2^2 + c_3 x_2^3.
\end{array}$$

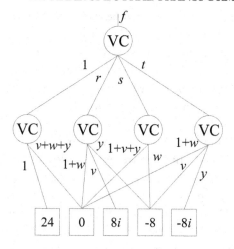

Figure 4.12: VCDD for f in Example 62.

4.8   HAAR SPECTRAL TRANSFORM DECISION DIAGRAMS

In the previous section, we consider decision diagrams defined with respect to the Fourier transforms on finite groups. These diagrams represent functions in the form of Fourier-series expressions. As noted in Chapter 3, the Haar functions are another orthogonal basis used efficiently in representations of discrete functions, in particular, binary and multiple-valued logic functions. Moreover, for some classes of logic functions Haar series provide compact representations in terms of the number of nodes count. For examples of such functions, see [82], [84], [99].

This was a motivation to define decision diagrams with respect to the Haar transform, the Haar spectral transform decision diagrams (HSTDDs), since they represent functions in terms of Haar-series and their generalizations. Thus, it is expected that functions with simple Haar and Haar-like spectra in the number of nodes count, will have simple HSTDDs in terms of the number of nodes or the number of paths.

The HSTDDs for multiple-valued functions are a multiple-valued counterpart of the Haar spectral transform decision diagrams (HSTDDs) for binary-valued functions [199], [218]. It should be noted that these diagrams are different from Haar spectral diagrams (HSDs) introduced in [75], [76]. The difference is that HSDs are Multi-terminal Binary Decision Diagrams for the function f to which the Haar spectral coefficients for f are also represented by attaching their values at appropriately determined positions. Thus, HSDs are MTBDDs for f with attached Haar spectral coefficients. The good feature of HSDs is that they represent Haar coefficients for f at the MTBDD for f which means there is no extra memory required compared to that for the MTBDD for f except for the negligible amount necessary to store the values of the coefficients. The HSD do not represent f in terms of the Haar series for f, and therefore, they cannot take advantage of these representations in the cases when the Haar spectrum is simpler to be represented by a decision diagram than the function itself. Moreover, there is no possibility of using the method for the reduction of Haar series to further reduce the related decision diagrams. With these considerations as a motivation, the Haar spectral transform decision diagrams were introduced and their applications discussed. For a review of these results and related references we refer to [199]. A generalization of Haar transforms and HSTDDs to multiple-valued functions was given in [220].

In the case of binary functions, HSTDDs can be viewed as hybrid diagrams between BDDs and WDDs, in the sense that Haar Spectral Transform Decision Diagrams (HSTDDs) consists of both Shannon and Walsh nodes. In the corresponding decision tree, nodes are distributed such that the Walsh nodes are situated as the leftmost nodes at each level in the diagram including the root node, while all other nodes are the Shannon nodes. This interpretation permits a straightforward extension to the HSTDD for multiple-valued functions by using nodes in MDD and Vilenkin-Chrestenson decision diagrams. In the general case, besides Fourier transform over the complex field, we can use nodes defined with respect to Fourier or Fourier-like transforms defined over finite fields.

Consider the space $P(C_p)$ of functions on the cyclic group C_p taking values in a field P that can be either the complex field C or a finite field. In $P(C_p)$, consider as the basis Q the Fourier basis or a basis used to define a polynomial expressions for $f \in P(C_p)$. Table 4.1 shows examples of such bases for different choices of P and C_p. Note that such bases are used to define the positive Davio expansions in various decision diagrams [216].

Definition 34 *For $f \in P(C_p^n)$, Haar Spectral Transform Decision Trees (HSTDTs) are decision trees where the expansion rules at the leftmost nodes at each level, including the root node, are defined by a basis Q for the Fourier series or polynomial expressions in $P(C_p)$ and for other nodes by the identity transforms for functions in $P(C_p)$.*

Definition 34 determines an algorithm for the design of HSTDTs.

Algorithm 2 *(Design of HSTDT)*

1. *Given a MTDT in $P(C_p^n)$.*

2. *Determine the positive Davio (pD) expansion from the basic matrix for the Fourier transform or the polynomial expressions in $P(C_p)$.*

3. *In MTDD, assign the pD–expansion to the leftmost nodes and relabel the outgoing edges of these nodes.*

4. *Reorder the variables in the descending order.*

5. *Determine the columns of a $(p^n \times p^n)$ matrix $\mathbf{Q}$ as product of labels at the edges.*

6. *Calculate the values of constant nodes by using the $\mathbf{Q}^{-1}$ inverse for $\mathbf{Q}$ over P.*

End of Algorithm.

Domain	Range	$\mathbf{Q}$
		Table 4.1: Bases in $P(C_p)$
C_2	C	$\mathbf{W}(1) = \begin{bmatrix} 1 & 1 \\ 1 & -1 \end{bmatrix}$
C_2	$GF(2)$	$\mathbf{R}(1) = \begin{bmatrix} 1 & 0 \\ 1 & 1 \end{bmatrix}$
C_3	C	$\mathbf{VC}_3(1) = \begin{bmatrix} 1 & 1 & 1 \\ 1 & e_1 & e_2 \\ 1 & e_2 & e_1 \end{bmatrix}$
C_3	$GF(3)$	$\mathbf{G}_{3GF}(1) = \begin{bmatrix} 1 & 0 & 0 \\ 1 & 1 & 1 \\ 1 & 2 & 1 \end{bmatrix}$
C_4	C	$\mathbf{VC}_4(1) = \begin{bmatrix} 1 & 1 & 1 & 1 \\ 1 & i & -1 & -i \\ 1 & -1 & 1 & -1 \\ 1 & -i & -1 & i \end{bmatrix}$
C_4	$GF(4)$	$\mathbf{G}_{4GF}(1) = \begin{bmatrix} 1 & 0 & 0 & 0 \\ 1 & 1 & 1 & 1 \\ 1 & 2 & 3 & 1 \\ 1 & 3 & 2 & 1 \end{bmatrix}$

Example 63 *Figrure 4.13 shows the HSTDT for ternary functions of two variables. The left most nodes are defined by the GF-expressions for single variable functions derived from the GF-transform matrix for* $p = 3$

$$\mathbf{G}_{3GF}(1) = \begin{bmatrix} 1 & 0 & 0 \\ 0 & 2 & 1 \\ 2 & 2 & 2 \end{bmatrix}.$$

Thus, each single variable function for $p =$ can be represented as

$$f = \begin{bmatrix} 1 & x_i & x_i^2 \end{bmatrix} \mathbf{G}_{3GF}(1)\mathbf{F},$$

where $\mathbf{F} = [f_0, f_1, f_2]^T$. Therefore,

$$f = f_0 \oplus x_i(2f_1 \oplus f_2) \oplus x_i^2(2f_0 \oplus 2f_1 \oplus 2f_2).$$

This expansion will be applied at the root node, to which the variable x_1 is assigned, and the leftmost node at the level for x_2. At all other nodes the generalized Shannon expansion rule is used. The decision tree defined by this assignment of nodes defines in turn the Haar transform for functions on C_3^2 over $GF(3)$. Conversely, we say that the HSTDD is defined in terms of a transform whose basis functions are obtained by multiplying labels at the edges in the decision trees as

$$1, x_2, x_2^2, x_1 J_0(x_2), x_1 J_1(x_2), x_1 J_2(x_2), x_1^2 J_0(x_2), x_1^2 J_1(x_2), x_1^2 J_2(x_2).$$

In matrix notation, these basis functions are written as columns of the matrix:

$$\mathbf{H}_{GF(3)}(2) = \begin{bmatrix} 1 & 0 & 0 & 0 & 0 & 0 & 0 & 0 & 0 \\ 1 & 1 & 1 & 0 & 0 & 0 & 0 & 0 & 0 \\ 1 & 2 & 1 & 0 & 0 & 0 & 0 & 0 & 0 \\ 1 & 0 & 0 & 1 & 0 & 0 & 1 & 0 & 0 \\ 1 & 1 & 1 & 0 & 1 & 0 & 0 & 1 & 0 \\ 1 & 2 & 1 & 0 & 0 & 1 & 0 & 0 & 1 \\ 1 & 0 & 0 & 2 & 0 & 0 & 1 & 0 & 0 \\ 1 & 1 & 1 & 0 & 2 & 0 & 0 & 1 & 0 \\ 1 & 2 & 1 & 0 & 0 & 2 & 0 & 0 & 1 \end{bmatrix}.$$

The matrix inverse to that is used to compute the corresponding Haar coefficients that are used as values of constant nodes in the considered HSTDT as:

$$\mathbf{H}_{GF(3)}^{-1}(2) = \begin{bmatrix} 1 & 0 & 0 & 0 & 0 & 0 & 0 & 0 & 0 \\ 0 & 2 & 1 & 0 & 0 & 0 & 0 & 0 & 0 \\ 2 & 2 & 2 & 0 & 0 & 0 & 0 & 0 & 0 \\ 0 & 0 & 0 & 2 & 0 & 0 & 1 & 0 & 0 \\ 0 & 0 & 0 & 0 & 2 & 0 & 0 & 1 & 0 \\ 0 & 0 & 0 & 0 & 0 & 2 & 0 & 0 & 1 \\ 2 & 0 & 0 & 2 & 0 & 0 & 2 & 0 & 0 \\ 0 & 2 & 0 & 0 & 2 & 0 & 0 & 2 & 0 \\ 0 & 0 & 2 & 0 & 0 & 2 & 0 & 0 & 2 \end{bmatrix}.$$

For instance, for the function f in Example 56 determined by the function vector $\mathbf{F} = [012120201]^T$, the Haar spectrum over $GF(3)$ is

$$\mathbf{S}_f = \mathbf{H}_{GF(3)}^{-1}(2)\mathbf{F} = [0, 1, 0, 1, 1, 1, 0, 0, 0]^T.$$

Therefore, by using this vector as the values of constant nodes in a HSTDT, this decision tree can be reduced into the HDTDD in Fig. 4.14. This decision diagram represents the given function in the form of a polynomial defined as

$$f = x_2 \oplus x_1 J_0(x_2) \oplus x_1 J_1(x_2) \oplus x_1 J_2(x_2) = x_2 \oplus x_1.$$

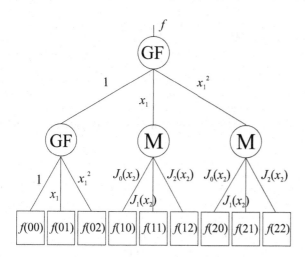

Figure 4.13: HSTDT for f in Example 63.

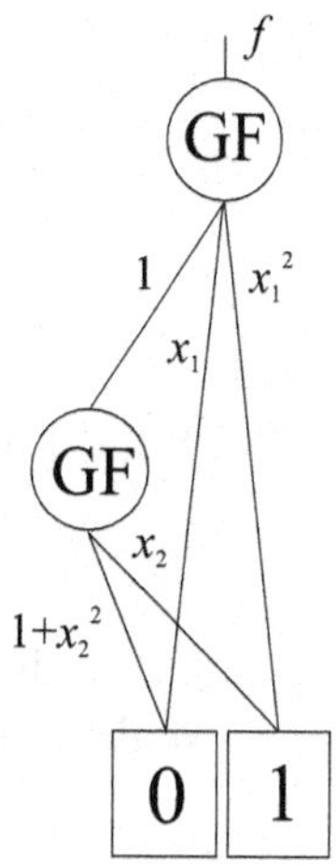

Figure 4.14: HSTDD for f in Example 63.

4.8.1 HSTDD RELATED TO THE RMF-TRANSFORM

In this section, we introduce these concepts on the example of the HSTDDs derived from the RMF-transform, since these diagrams are not previously discussed in the literature.

Recall that the basic Reed-Muller-Fourier (RMF) transform matrix for quaternary functions $\mathbf{R}_4(1)$ and the inverse matrix defining the basis functions in terms of which this transform is defined $\mathbf{X}_{4RMF}(1)$ are

$$\mathbf{R}_{4RMF}(1) = 3 \begin{bmatrix} 1 & 0 & 0 & 0 \\ 1 & 3 & 0 & 0 \\ 1 & 2 & 1 & 0 \\ 1 & 1 & 3 & 3 \end{bmatrix}, \quad \mathbf{X}_{4RMF}(1) = \begin{bmatrix} 3 & 0 & 0 & 0 \\ 3 & 1 & 0 & 0 \\ 3 & 2 & 3 & 0 \\ 3 & 3 & 1 & 1 \end{bmatrix}.$$

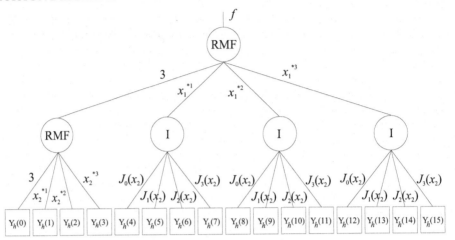

Figure 4.15: HSDT for quaternary functions of two variables defined with respect to the RMF-expressions.

When written symbolically $\mathbf{X}_{4RMF}(1) = [3, x_i^{*1}, x_i^{*2}, x_i^{*3}]$. Therefore, from $\mathbf{R}(1)$ and $\mathbf{X}_{4RMF}(1)$, we derive the expansion rule

$$
\begin{aligned}
f &= \begin{bmatrix} 3 & x_i^{*1} & x_i^{*2} & x_i^{*3} \end{bmatrix}
\begin{bmatrix} 1 & 0 & 0 & 0 \\ 1 & 3 & 0 & 0 \\ 1 & 2 & 1 & 0 \\ 1 & 1 & 3 & 3 \end{bmatrix}
\begin{bmatrix} f(0) \\ f(1) \\ f(2) \\ f(3) \end{bmatrix} \\[2mm]
&= \begin{bmatrix} 3 & x_i^{*1} & x_i^{*2} & x_i^{*3} \end{bmatrix}
\begin{bmatrix} f(0) \\ f(0) \oplus 3f(1) \\ f(0) \oplus 2f(1) \oplus f(2) \\ f(0) \oplus f(1) \oplus 3f(2) \oplus 3f(3) \end{bmatrix} \\[2mm]
&= f(0) \oplus x_i(f(0) \oplus 3f(1)) \oplus x_i^{*2}(f(0) \oplus 2f(1) \oplus f(2)) \\
&\quad \oplus x_i^{*3}(f(0) \oplus f(1) \oplus f(2) \oplus f(3)).
\end{aligned}
\tag{4.13}
$$

Definition 35 *The Haar spectral transform decision trees (HSTDTs) are defined as the decision trees to whose root node and the left-most nodes at each level the expansion rule derived from the basic spectral transform is assigned, while to all other nodes the generalized Shannon expansion rule is assigned. The Haar spectral transform decision diagrams (HSTDDs) are defined by the reduction of the HSDTs by removing redundant information that can be done by using the correspondingly extension of the generalized BDD reduction rules to the multiple-valued case.*

Example 64 *Figure 4.15 shows the HSDTD defined in terms of the RMF-transform for quaternary functions of n = 2 variables. In this decision tree, the products of labels at the edges along paths form the* *root node to the constant nodes correspond to the basis functions of the Haar-RMF-transform in Example 36 specified by (2.12). The constant nodes show the values of the corresponding spectral coefficients determined by using the Haar-RMF-transform matrix defined by (2.13),*

$$\mathbf{S}_{f,hY} = \mathbf{Y}_{h,4RMF}(2)\mathbf{F},$$

where $\mathbf{F}$ *is the function vector of the function to be represented.*

4.9   EDGE-VALUED DECISION DIAGRAMS

In representations of discrete functions by MTBDDs or MDDs, it may happen that a function has many different values resulting in many constant nodes. This usually immediately means a large size of the diagram, since branching at many places is necessary to provide paths from the root node to each constant node. Therefore, such diagrams are inefficient and, as an attempt to resolve this problem, the Edge-valued decision diagrams (EVDDs) were introduced for binary-valued logic functions, and thus called Edge-valued binary decision diagrams (EVBDDs) [103]. The main idea behind these diagrams is to use additive or multiplicative factors that may appear in the values of constant nodes. These factors are assigned to the edges connecting non-terminal nodes and the constant nodes are replaced by a single node usually showing the value 0 or 1 depending on the factors used as the attributes at the edges and related normalization methods that are applied to ensure canonicity of representations.

These modifications are formalized through the corresponding redefinition of the decomposition rules used to assign functions to edge-valued decision diagrams. In this way, a variety of different edge-valued decision diagrams, i.e., decision diagrams with attributed edges was defined starting by [103].

In spectral interpretation of decision diagrams [195], a given function f is assigned to a decision tree by performing a spectral transform to the function represented, which in terms of decision trees can be expressed as a recursive application of various decomposition rules at the nodes of decision trees [170]. Due to this interpretation, various decision diagrams defined with respect to different decomposition rules are uniformly viewed as examples of Spectral transform decision diagrams (STDDs) defined with respect to different spectral transforms. The spectral interpretation extends to edge-valued decision diagrams showing that the main difference with respect to the other classes of diagrams is in the way how the related spectral transforms are performed to assign a function to the decision diagram. This topics will be discussed below.

A spectral transform is specified by the set of selected basis functions Q. It should be assured that the basis functions in terms of which a spectral transform is defined correspond in some way to the recursive structure of decision trees [195].

Application of the decomposition rules at nodes in a level of a decision tree corresponds to the implementation of steps of fast computation algorithms (FFT-like) for spectral transforms used in the definition of the corresponding decision tree.

Table 4.2: Discrete transforms and decision diagrams for MV functions

Transform	Decision diagram
Identical MDD [229]	MD-SV [118]
MTMDD [116]	QDD [169]
GF-transform	GFDD [186]
	(Pseudo-QDD) [169]
RMF-transform	Reed-Muller-Fourier DD [186]
Partial GF	EVGFDD [210]
Partial RMF	EVRMFDD [187]

In EVDDs, the attributes at the edges can be determined by referring to the partial spectral transforms defined in terms of steps in FFT-like algorithms for calculation of related spectral transforms. In EVDDs with additive weights, the attributes to the edges are determined as values of some subsets of partial transform coefficients of the functions represented. The multiplicative weights are determined as common factors in intermediate values of spectral coefficients after performing steps of FFT-like algorithms. Steps in FFT-like algorithms are related to the levels in decision diagrams. Therefore, the difference between EVDDs with additive and multiplicative weights is in the way of performing steps of FFT-like algorithms. For additive weights, all steps are performed over f. For multiplicative weights, steps are performed recursively, just the first step over f while inputs of other steps are outputs of the previous steps. In steps of FFT, application of Shannon nodes corresponds to no processing, and due to that permits to extract additive weights together with multiplicative weights for edges in EVDDs derived from the corresponding Kronecker decision diagrams. Another difference with respect to other decision diagrams and also among different EVBDDs is that in EVDDs with additive weights, the constant nodes are set to zero, while in these with multiplicative weights, the constant nodes show the values of the last factors in the decomposition of the related spectral coefficients.

In particular, EVBDDs [103], [244] and *BMDDs [21] are variants of decision diagrams related to the Arithmetic transform [185], [195]. K*BMDs [35] are a generalization of *BMDs by using also the integer counterparts of Shannon nodes, thus, relating to the Kronecker transforms, and therefore, having both additive and multiplicative weights.

Figure 4.16 explains assignment of a given function f to edge-valued decision diagrams with additive and multiplicative weights at the edges. For simplicity, we show decision diagrams with nodes having two outgoing edges. However, the same directly applies to decision diagrams for multiple-valued functions if we use nodes with p outgoing edges and related spectral transforms for multiple-valued functions, thus, transforms described by $(p^n \times p^n)$ transform matrices. In decision diagrams, we usually select Kronecker transforms, in which case the transform matrix is Kronecker product representable, i.e., $\mathbf{Q} = \otimes_{i=1}^{n} \mathbf{Q}_i$, where $\mathbf{Q}_i$ is the basic transform matrix [195]. In HSTDD, we use the layer-Kronecker product representable matrices.

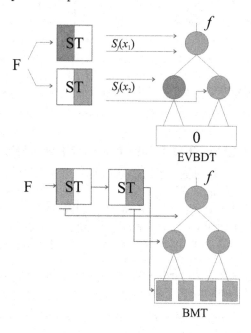

Figure 4.16: Assignment of functions to edge-valued decision diagrams.

Table 4.2 shows a classification of EVDDs for multiple-valued functions with respect to spectral transform in terms of which they are defined. In this table, MD-SV are a special class of EVDDs defined with respect to the identity transform and by exploiting symmetry properties of multiple-valued functions [118].

4.9.1   PARTIAL REED-MULLER-FOURIER TRANSFORMS

In this section, we consider the edge-valued decision diagrams based on the RMF-representations. The presentation is given on the example of multiple-valued functions in $GF(4)$. Extensions to multiple-valued functions for an arbitrary number of function values, i.e., in different Galois fields and related algebraic structures as the multiple-valued Gibbs algebra is straightforward [67], [69].

By the analogy to the partial Reed-Muller transforms for switching functions, the partial Reed-Muller-Fourier transforms for multiple-valued functions are defined through the Good-Thomas factorization of the RMF-matrix (see Chapter 5).

Definition 36 *The partial Reed-Muller-Fourier spectrum of a function* f *given by the truth-vector* $\mathbf{F} = [f(0), \ldots, f(4^n - 1)]^T$ *is defined by*

$$\mathbf{S}_{f,4RMF} = \mathbf{C}_{4RMF}\mathbf{F},$$

where $\mathbf{C}_{4RMF}$ *is derived by the factorization of* $\mathbf{R}_4(n)$ *and is given by*

$$\mathbf{C}_{4RMF} = \bigotimes_{j=1}^{n} \mathbf{X}_4(1), \quad \mathbf{X}_4(1) = \begin{cases} \mathbf{R}_4(1), & i = j, \\ \mathbf{I}_4(1), & i \neq j, \end{cases}$$

where $\mathbf{I}_4(1)$ *is the identity matrix of order* 4 *and* $\mathbf{R}_{4RMF}(1) = \begin{bmatrix} 1 & 0 & 0 & 0 \\ 1 & 3 & 0 & 0 \\ 1 & 2 & 1 & 0 \\ 1 & 1 & 3 & 3 \end{bmatrix}.$

4.9.2   EDGE-VALUED REED-MULLER-FOURIER DECISION DIAGRAMS

EVBDDs are introduced as the decision diagrams derived through the reduction of the decision tree defined in terms of the algebraic function $x_i v_1 + f_0 + x_i(f_1 - f_0)$ [103]. Edge-valued functional decision diagrams (EVFDDs) are defined by using the same function, but over $GF(2)$ $x_i v_1 \oplus f_0 \oplus x_i(f_0 \oplus f_1)$ [187], [210]. We define the edge-valued Reed-Muller-Fourier decision diagrams (EVRMFDDs) to represent multiple-valued functions in the same way.

Definition 37 *An edge-valued Reed-Muller-Fourier decision diagram is a pair* $(c, \mathbf{f})$, *where* c *is a constant and* $\mathbf{f}$ *is a directed acyclic graph consisting of two different sets of nodes*

1. *A single terminal or constant node representing the value 0 and, thus denoted by 0.*

2. *A non-terminal node* v *described by* $\langle var(v), \phi_0(v), \phi_1(v), \phi_2(v), \phi_3(v), r_0, r_1, r_2, r_3 \rangle$, *where* $var(v) \in \{x_1, \dots, x_n\}$, $\phi_0(v), \phi_1(v), \phi_2(v), \phi_3(v)$ *are EVRMFDDs that represent the subfunctions* $f_0 = f(x_i = 0)$, $f_1 = f(x_i = 1)$, $f_2 = f(x_i = 2)$, $f_3 = f(x_i = 3)$ *and* r_0, r_1, r_2, r_3 *are the weighting coefficients assigned to the outgoing edges of* v *corresponding to* f_0, f_1, f_2, f_3.

As in EVBDDs, c is the value of f at the point 0 and the coefficient assigned to the outgoing branch corresponding to f_0 is set to zero to provide the uniqueness of the representation of f by an EVRMFDD.

Definition 38 *A multiple-valued function* f *is attached to a EVRMFDD through the following rules.*

1. *If* q *is a constant node with the value* w, *then it represent the function* $f_q = w$.

2. *If* q *is a non-terminal node related to the variable* x_i, $i = 2, \dots, n$, *then*

$$\begin{aligned} f_{x_i} &= x_i r_i^1 \oplus x_i^{*2} r_i^2 \oplus x_i^{*3} r_i^3 \oplus f_0 \oplus x_i(f_0 \oplus 3f_1) \\ &\oplus x_i^{*2}(f_0 \oplus 2f_1 \oplus f_2) \oplus x_i^{*3}(f_0 \oplus f_1 \oplus 3f_2 \oplus 3f_3), \end{aligned}$$

and for the root node

$$\begin{aligned} f_{x_1} &= x_1 r_1^1 \oplus x_1^{*2} r_1^2 \oplus x_1^{*3} r_1^3 \oplus f_0 \oplus x_1(3f_0 \oplus f_1) \\ &\oplus x_1^{*2}(3f_0 \oplus 2f_1 \oplus 3f_2) \oplus x_1^{*3}(3f_0 \oplus 3f_1 \oplus f_2 \oplus f_3). \end{aligned}$$

To do the optimization, the reduction of EVRMFDD is done in the same way as with EVBDDs [103]. To estimate the complexity of an EVRMFDD it is convenient to consider it as the edge-valued decision diagram derived by the reduction of the corresponding edge-valued decision tree. In that case, it can be shown, by the analogy to the spectral interpretation of EVBDDs [185] and EVFDDs [208], that the complexity of EVRMFDDs depends on the structure of vectors representing the partial Reed-Muller-Fourier transforms of multiple-valued functions.

Example 65 *Figure 4.17 shows the edge-valued Reed-Muller-Fourier decision tree in $GF(4)$ for $n = 2$. From the definition of EVRMFDDs, it follows that this tree represents the function in the form of a RMF-polynomial with reordered coefficients. For that reordering the complexity of an EVRMFDD depends on the structure of the partial RMF-transforms spectra, since the values assigned to the edges are determined as the particular coefficients of the partial RMF-transforms. In Fig. 4.17, we show the rows of the RMF-transform matrix in terms of which the coefficients of the partial RMF-transforms used as attributes at the edges are computed.*

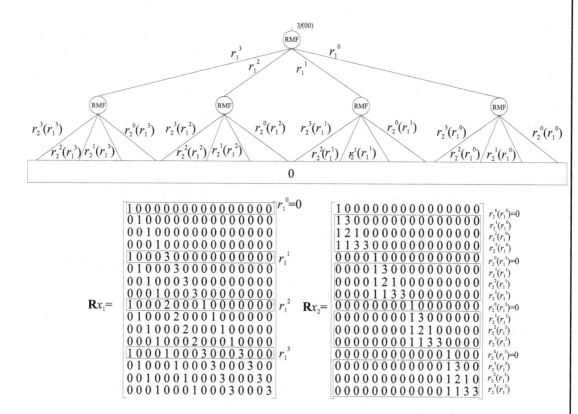

Figure 4.17: Edge-valued Reed-Muller-Fourier decision tree for $p = 4$ and $n = 2$.

Example 66 *Figure 4.18 shows the reduced EVRMFDD of a two-variable function f in $GF(4)$ given by the truth-vector* $\mathbf{F} = [0, 0, 0, 0, 2, 2, 2, 2, 1, 3, 0, 2, 2, 0, 3, 1]^T$. *From definition of EVRMFDDs, this decision diagram represents the function f in the form of the RMF-expressions as*

$$f = x_1^{*2} \oplus x_1^{*3} \oplus 2x_1 \odot x_2 \oplus x_1 \odot x_2^{*2} \oplus 2x_1 \odot x_2^{*3} \oplus 2x_1^{*2} \odot x_2^{*2}.$$

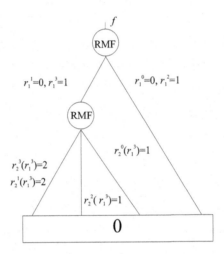

Figure 4.18: Edge-valued Reed-Muller-Fourier decision diagram for f in Example 66.

In the same way, we can define many different edge-valued decision diagrams by using various partial spectral transforms defined by steps of the corresponding fast algorithms as will be discussed below. For example, the Edge-valued Galois field decision diagrams were introduced in [210].

4.10   CONSTRUCTION OF EVDDS

From the spectral interpretation of decision diagrams, edge-valued decision diagrams are actually a different notation of ordinary decision diagrams. Therefore, for each decision diagram, an edge-valued version can be assigned by using the following algorithms.

Algorithm 3 *(EVDDs with additive weights)*

1. *Given a multiple-valued function f of n variables.*

2. *Select a Kronecker spectral transform with respect to a basis Q, ST(Q), in terms of which a decision diagram is required.*

3. *For $i = 1$ to n, apply the i-th step of an FFT-like algorithm for ST(Q) to f. The result is denoted as $S_{i,f}$.*

4. *Determine common factors q_i in $S_{i,f}$.*

5. *Assign factors q_i as weight coefficients to the left outgoing edges of a decision tree.*

6. *Set the constant nodes of the DT to zero and perform the reduction by deleting isomorphic subtrees.*

Algorithm 4 *(EVDDs with multiplicative weights)*

1. *Given a multiple-valued function f of n variables.*

2. *Select a Kronecker spectral transform with respect to a basis Q, $ST(Q)$, in terms of which a decision diagram is required.*

3. *Perform an FFT-like algorithm of n steps to f and calculate the spectrum S_f with respect to $ST(Q)$.*

4. *Factorize the spectral coefficients in S_f into product of n factors, attempting to have equal factors at the same positions.*

5. *If the factors at the i-th position where $i = 1, \ldots n - 1$ are equal, assign them as weighted coefficients at the edges at the i-th level in a decision tree.*

6. *The n-th factors show as values of constant nodes in the DT.*

7. *Perform a reduction of the decision tree by deleting isomorphic subtrees.*

In EVDDs constructed by these algorithms, labels at the edges are determined as functions $\phi(i)$ describing columns of matrices inverse to the basic transform matrices used in the definition of the i-th step of FFT-like algorithms. From an EVDD, we read the function f that is represented by following the labels at the edges and taking into account the assigned weights in the same way as in the case of ordinary decision diagrams or EVDDs for binary functions. More details about these decision diagrams can be found in [170] for binary and [187], [196], and [210] for MV functions.

4.10.1 EFFICIENCY OF EVDDS

The reduction of a decision tree into a decision diagram is possible due to the existence of constant subvectors or mutually equal subvectors in the vector $\mathbf{V}$ of values of constant nodes in the decision tree. In decision diagrams, a constant subvector is represented by a single constant node, while equal subvectors result in isomorphic subtrees that can be represented by a single subtree. Efficiency of EVDDs in representation of some classes of functions is explained by the following remark.

Remark 1 *In reduction of EVDDs, besides values of constant and isomorphic subvectors in $\mathbf{V}$, common factors, additive or multiplicative, in elements of $\mathbf{V}$ are taken into account.*

From this remark, we determine the following characterization of classes of functions where EVDDs can be efficient in the number of nodes count.

Remark 2 *Functions whose spectral coefficients with respect to a basis Q have many common multiplicative factors can be efficiently represented by EVDD(Q)s with multiplicative weights.*

Functions whose partial spectral coefficients with respect to a basis Q have many equal additive factors, can be efficiently represented by EVDD(Q)s with additive weights.

Besides compactness, complexity of manipulation with decision diagrams is another important issue in their applications. In that respect, it should be noted that the complexity of manipulation with decision diagrams constructed by the algorithms proposed is the same, or a least not greater than that of the manipulation with other decision diagrams for MV functions provided that implementation is based on case structures [36], [37], and exploiting basic principles in programming of decision diagrams for MV functions [117].

4.10.2   ILLUSTRATIVE EXAMPLES

In this section, we study the application of the proposed algorithms for constructing two classes of edge-valued decision diagrams with respect to three different spectral transforms. These decision diagrams are multiple-valued analogs of EVBDDs [103] and *BMDs [21]. Since different edge-valued decision diagrams can be defined for different spectral transforms Q, we denote them as EVMDD(Q)s and *MMDs, with M standing for multi-valued functions as in Multiple-place decision diagrams (MDDs) [177], and Q specifying the transform used in definition of the particular decision diagram.

The following example shows an EVRMFDD for a concrete quaternary function for $n = 2$ variables.

Example 67 *(EVRMFDD)*

Figure 4.18 shows the reduced EVRMFDD of a two-variable function f in $GF(4)$ given by the truth-vector $\mathbf{F} = [0, 0, 0, 0, 2, 2, 2, 2, 1, 3, 0, 2, 2, 0, 3, 1]^T$. This EVRMFDD represents the function f as $f = x_1^{*2} \oplus x_1^{*3} \oplus 2x_1x_2 \oplus x_1x_2^{*2} \oplus 2x_1x_2^{*3} \oplus 2x_1^{*2}x_2^{*2}.$

The following example illustrates application of the second algorithm to the construction of edge valued version of Multiple-valued moment diagrams (MVMDs) defined with respect to the arithmetic transforms. MVMDs and their edge-valued version *MVMDs are generalizations of BMDs and *BMDs to multiple-valued functions. The arithmetic transform is a particular case of spectral transforms, and it is selected for this example since the same transform is used in the definition of BMDs. It should be noted that MVMDs and *MVMDs can be defined with respect to different spectral transforms with real-valued or even complex-valued coefficients. Thus, these decision diagrams constitute a broad family of decision diagrams that can be defined for different choices of spectral transforms. Since these decision diagrams are defined with respect to real-valued or complex-valued spectral transforms, MVMDs and *MVDs can be used also to represent integer or complex-valued functions.

Example 68 *(*MVMDs for Q = Arithmetic transform)*

Assume that Q is the arithmetic transform applied to quaternary functions. Thus, it is described by the *basic transform matrix*

$$\mathbf{Q}(1) = \begin{bmatrix} 1 & 0 & 0 & 0 \\ -1 & 1 & 0 & 0 \\ -1 & 0 & 1 & 0 \\ 1 & -1 & -1 & 1 \end{bmatrix}.$$

Consider a three-variable five-output quaternary function f that can be represented by the integer equivalent function derived by summation of outputs with weights 4^i, $i = 0, 1, 2, 3$, given by the vector of function values

$$\begin{aligned} \mathbf{F} = \quad & [88, 116, 266, 181, 66, 73, 180, 112, 116, 104, 204, 140, 52, 60, 118, 82, \\ & 12, 44, 100, 72, 58, 52, 78, 50, 48, 52, 88, 60, 60, 48, 66, 42, \\ & 176, 140, 254, 161, 118, 85, 164, 102, 156, 112, 184, 120, 72, 60, 108, 72, \\ & 84, 68, 104, 68, 100, 64, 80, 48, 76, 52, 76, 48, 72, 48, 60, 36]^T. \end{aligned}$$

The representation of this function by an MDD [177] requires 21 non-terminal nodes and 42 constant nodes. The arithmetic spectrum is given by the vector of spectral coefficients

$$\begin{aligned} \mathbf{S}_f = \quad & [-2, -1, 0, 1, -2, -1, 0, 1, -2, -1, 0, 1, -2, -1, -1, 1, \\ & -8, -8, -4, -4, -8, -8, -4, -4, -12, -6, 0, 6, -12, -6, 0, 6, \\ & 0, 0, 3, 3, -9, -9, 18, 18, 12, 12, 24, 36, -24, -24, 12, 36, \\ & -16, -8, 0, 8, 4, 4, 8, 12, -8, -8, 4, 12, 0, 12, 24, 36]^T. \end{aligned}$$

This vector can be used to represent the given function f by an MVMD which is a multiple-valued analog of BMDs [21] having the root node at the level for x_1, 3 non-terminal nodes at the level for x_2, and 11 non-terminal nodes at the level for x_3. There are 16 constant nodes. Notice that five constant nodes can be saved if nodes with negated edges are allowed.

We factorize this vector as

$$
\begin{aligned}
\mathbf{S}_f = [&2 \cdot 2 \cdot (-2), 2 \cdot 2 \cdot (-1), 2 \cdot 2 \cdot 0, 2 \cdot 2 \cdot 1, \\
&2 \cdot 2 \cdot (-2), 2 \cdot 2 \cdot (-1), 2 \cdot 2 \cdot 0, 2 \cdot 2 \cdot 1, \\
&2 \cdot 2 \cdot (-2), 2 \cdot 2 \cdot (-1), 2 \cdot 2 \cdot 0, 2 \cdot 2 \cdot 1, \\
&2 \cdot 2 \cdot (-2), 2 \cdot 2 \cdot (-1), 2 \cdot 2 \cdot 0, 2 \cdot 2 \cdot 1, \\
&2 \cdot 2 \cdot (-2), 2 \cdot 2 \cdot (-2), 2 \cdot 2 \cdot (-1), 2 \cdot 2 \cdot (-1), \\
&2 \cdot 2 \cdot (-2), 2 \cdot 2 \cdot (-2), 2 \cdot 2 \cdot (-1), 2 \cdot 2 \cdot (-1), \\
&2 \cdot 3 \cdot (-2), 2 \cdot 3 \cdot (-1), 2 \cdot 3 \cdot 0, 2 \cdot 3 \cdot 1, \\
&2 \cdot 3 \cdot (-2), 2 \cdot 3 \cdot (-1), 2 \cdot 3 \cdot 0, 2 \cdot 3 \cdot 1, \\
&3 \cdot 1 \cdot 0, 3 \cdot 1 \cdot 0, 3 \cdot 1 \cdot 1, 3 \cdot 1 \cdot 1, \\
&3 \cdot 3 \cdot (-1), 3 \cdot 3 \cdot (-1), 3 \cdot 3 \cdot 2, 3 \cdot 3 \cdot 2, \\
&3 \cdot 4 \cdot 1, 3 \cdot 4 \cdot 1, 3 \cdot 3 \cdot 2, 3 \cdot 4 \cdot 3, \\
&3 \cdot 4 \cdot (-2), 3 \cdot 4 \cdot (-2), 3 \cdot 4 \cdot 1, 3 \cdot 4 \cdot 3, \\
&4 \cdot 2 \cdot (-2), 4 \cdot 2 \cdot (-1), 4 \cdot 2 \cdot 0, 4 \cdot 2 \cdot 1, \\
&4 \cdot 1 \cdot 1, 4 \cdot 1 \cdot 1, 4 \cdot 1 \cdot 2, 4 \cdot 1 \cdot 3, \\
&4 \cdot 1 \cdot (-2), 4 \cdot 1 \cdot (-2), 4 \cdot 1 \cdot 1, 4 \cdot 1 \cdot 3, \\
&4 \cdot 3 \cdot 0, 4 \cdot 3 \cdot 1, 4 \cdot 3 \cdot 2, 4 \cdot 3 \cdot 3]^T
\end{aligned}
$$

and due to this, we can represent f by an $*MVMD$, which is a multiple-valued analog to $*BMDs$ [21] with the root node at the level for x_1, 3 non-terminal nodes at the level for x_2, and 8 non-terminal nodes at the level for x_3. First and second factors are weights at the edges at the levels for x_1 and x_2, respectively. There are 6 constant nodes showing values of the third factors, thus, -2,-1,0,1,2, and 3. Therefore, the reduction achieved by transferring from MVMD to $*MVMD$ consists of 3 non-terminal nodes at the level for x_3 and 11 constant nodes. Figure 4.19 shows this $*MVMD$ for the considered function f. In this $*MVMD$, labels at the edges are 1, w_2, w_1, and $w_1 w_2$, where w_1 and w_2 are binary valued variables. This follows from columns of the matrix $\mathbf{Q}^{-1}(1)$, as noted above.

4.11   CONSTRUCTION OF TRANSFORMS FROM DECISION DIAGRAMS

Spectral interpretation of decision diagrams states that a decision diagram represents the given function in the form of a spectral expression in terms of basis functions determined by the decomposition rules used to define the underlying decision tree. Therefore, given a decision tree, we can determine the basis functions in terms of which it is defined. It follows that each decision tree defines uniquely a spectral transform. By changing nodes in a decision tree, we can define new spectral transforms sharing the same structure and, therefore, also the properties of the initial spectral transform used in the definition of the considered decision tree. In particular, we can define various forms of Haar transforms by selecting different leftmost nodes in the MDTs and reading the product of labels at the edges as the corresponding basis functions for these transforms. In this sense, the following definition of Haar functions can be stated [220].

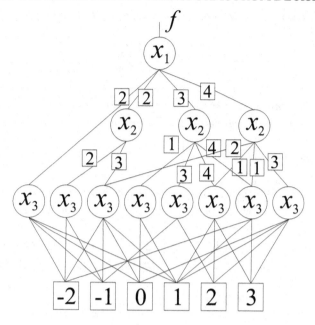

Figure 4.19: ∗MVMD for f in Example 68.

Definition 39 *Haar functions in $P(C_p^n)$, are functions described by products of labels at the edges along the paths from the root node to the constant nodes in the HSTDTs.*

If basis functions determined as specified by Definition 34 are written as columns of a matrix $\mathbf{Q}$, then the matrix $\mathbf{Q}^{-1}$ is the transform matrix for a Haar-like spectral transform.

In what follows, we review examples of Haar transforms derived from HSTDDs for different choices of domains C_p, ranges P, and basic matrices in Fourier and polynomial expression in $P(C_p^n)$.

4.11.1 VILENKIN-CHRESTENSON HAAR TRANSFORM

The matrix $\mathbf{W}(1)$ is the basic Walsh matrix, and the Walsh transform is the Fourier transform in $C(C_2^n)$. The Vilenkin-Chrestenson transform is the Fourier transform in $C(C_p^n)$ [98]. Therefore, a generalization of the HSTDDs to functions in $C(C_p^n)$ is derived by using this transform. The method will be explained by the following example.

Example 69 *For $p = 3$, the basic Vilenkin-Chrestenson matrix is given by*

$$\mathbf{VC}_3(1) = \begin{bmatrix} 1 & 1 & 1 \\ 1 & e_1 & e_2 \\ 1 & e_2 & e_1 \end{bmatrix},$$

where $e_1 = -\frac{1}{2}(1 - i\sqrt{3})$, and $e_2 = e_1^* = -\frac{1}{2}(1 + i\sqrt{3})$, where z^* denotes the complex-conjugate of z.

The Vilenkin-Chrestenson transform is defined by the transform matrix

$$\mathbf{VC}^{-1} = \mathbf{VC}^* = \bigotimes_{i=1}^{n} \mathbf{VC}_3(1).$$

Vilenkin-Chrestenson-Haar spectral transform DTs (VCHSTDTs) are defined as decision trees where the expansion rule for the leftmost nodes is determined by the matrix $\mathbf{VC}(1)$, and for the other nodes by the identity matrix $\mathbf{I}_3 = \begin{bmatrix} 1 & 0 & 0 \\ 0 & 1 & 0 \\ 0 & 0 & 1 \end{bmatrix}$.

In VCHSTDTs, the values of constant nodes are the Vilenkin-Chrestenson coefficients. The labels at the edges are determined by the analytic expression for columns of $\mathbf{VC}_3(1)$. Thus, they are $r_0 = 1$, $r_1 = 1 + vx + dx^2$, $r_2 = 1 + v^*x + d^*x^2$, where $v = -i\sqrt{3}$, and $d = -\frac{3}{2}(1 - i\sqrt{3})$.

Figure 4.20 shows VCHSTDT for $n = 2$. In this VCHSTDT, products of labels at the edges determine the Haar functions in $C(C_3^2)$ [98]. In the notation used in VCHSTDTs, they are

$$
\begin{aligned}
har_3(0) &= 1, \\
har_3(1) &= 1 + vx_1 + dx_1^2, \\
har_3(2) &= 1 + v^*x_1 + d^*x_1^2, \\
har_3(3) &= (1 + vx_2 + dx_2^2)J_0(x_1), \\
har_3(4) &= (1 + vx_2 + dx_2^2)J_1(x_1), \\
har_3(5) &= (1 + vx_2 + dx_2^2)J_2(x_1), \\
har_3(6) &= (1 + v^*x_2 + d^*x_2)J_0(x_1), \\
har_3(7) &= (1 + v^*x_2 + d^*x_2)J_1(x_1), \\
har_3(8) &= (1 + v^*x_2 + d^*x_2^2)J_2(x_1),
\end{aligned}
$$

where $J_i(x_j)$ are characteristic functions defined as $J_i(x_j) = 1$ for $x_j = i$, and $J_i(x_j) = 0$ for $x_j \neq i$. In Fig. 4.20, S_3 denotes the generalized Shannon expansion in $C(C_3)$ defined as

$$f = J_0(x_i)f_0 + J_1(x_i)f_1 + J_2(x_i)f_2,$$

where f_i, $i = 0, 1, 2$ are co-factors of f for $x_i \in 0, 1, 2$. The nodes labeled by $\mathbf{VC}_3$ are the positive Davio nodes representing the positive Davio expansion defined from $\mathbf{VC}_3(1)$ as

$$
\begin{aligned}
f &= 1 \cdot S_{f_0} + (1 + vx_i + dx_i^2)S_{f_1} + (1 + v^*x_i + d^*x_i^2)S_{f_2}, \\
&= 1 \cdot (f_0 + f_1 + f_2) + (1 + vx_i + dx_i^2)(f_0 + e_2 f_1 + e_1 f_2) \\
&\quad + (1 + v^*x_i + d^*x_i^2)(f_0 + e_1 f_1 + e_2 f_2).
\end{aligned}
$$

It should be noticed that the Vilenkin-Chrestenson-Haar (VCH) transform, up to reordering is equivalent to the Watari transform [72], [138], [139], [148], [245]. This means that the decision tree discussed above may be properly modified to generate Watari transforms.

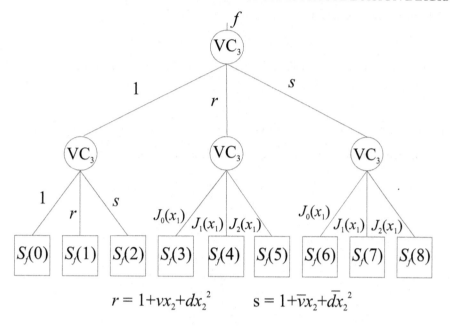

$$r = 1+vx_2+dx_2^2 \qquad s = 1+\bar{v}x_2+\bar{d}x_2^2$$

Figure 4.20: VCHSTDT for $n = 2$.

4.11.2 GALOIS FIELD HAAR TRANSFORMS

Even though Galois field Haar transforms have been suggested long ago [98], no development work seems to have been done. In what follows, we define HSTDDs and related Haar transforms derived from Galois field (GF) polynomial expressions. The method will be explained by the example of functions in $GF_3(C_3^2)$, which is the space of ternary functions of two variables, i.e., functions defined on C_3^2 and taking values in $GF(3)$.

For $P = GF(3)$, the basic functions in GF-expressions are defined by the GF-matrix $\mathbf{X}_{3GF}(1)$. The columns of this matrix can be expressed as 1, x, and x^2, where $x \in \{0, 1, 2\}$ and calculations are in $GF(3)$. The inverse matrix is

$$\mathbf{G}_{3GF}^{-1} = \begin{bmatrix} 1 & 0 & 0 \\ 0 & 2 & 1 \\ 2 & 2 & 2 \end{bmatrix}.$$

From this matrix, we determine the positive Davio expansion as

$$\begin{aligned} f &= 1 \cdot S_{f_0} \oplus x_i S_{f_1} \oplus x_i^2 S_{f_2} \\ &= 1 \cdot f_0 \oplus x_i(2f_1 \oplus f_2) \oplus x_i^2(2f_0 \oplus 2f_1 \oplus 2f_2), \end{aligned}$$

and define the related positive Davio nodes.

The Shannon expansions in $GF_3(C_3^n)$ are defined as

$$f = J_0(x_i) f_0 \oplus J_1(x_i) f_1 \oplus J_2(x_i) f_2.$$

Figure 4.21 shows GFHSTDT for $n = 2$. Product of labels in this GFHSTDT determine the Galois-Field-Haar functions in $GF_3(C_3^2)$.

1, x_1, x_1^2, $x_2 J_0(x_1)$, $x_2 J_1(x_1)$, $x_2 J_2(x_1)$ $x_2^2 J_0(x_1)$, $x_2^2 J_1(x_1)$, $x_2^2 J_2(x_1)$.

In matrix notation, these functions are written as columns of the matrix

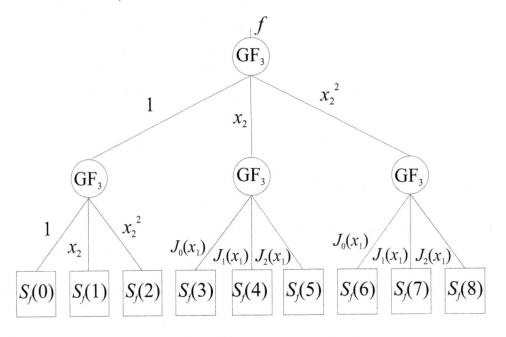

Figure 4.21: GFHSTDT for $n = 2$.

$$\mathbf{GFH}_3(2) = \begin{bmatrix} 1 & 0 & 0 & 0 & 0 & 0 & 0 & 0 & 0 \\ 1 & 0 & 0 & 1 & 0 & 0 & 1 & 0 & 0 \\ 1 & 0 & 0 & 2 & 0 & 0 & 1 & 0 & 0 \\ 1 & 1 & 1 & 0 & 0 & 0 & 0 & 0 & 0 \\ 1 & 1 & 1 & 0 & 1 & 0 & 0 & 1 & 0 \\ 1 & 1 & 1 & 0 & 2 & 0 & 0 & 1 & 0 \\ 1 & 2 & 1 & 0 & 0 & 0 & 0 & 0 & 0 \\ 1 & 2 & 1 & 0 & 0 & 1 & 0 & 0 & 1 \\ 1 & 2 & 1 & 0 & 0 & 2 & 0 & 0 & 1 \end{bmatrix}.$$

The inverse matrix is

$$
\mathbf{GFH_3(2)}^{-1} =
\begin{bmatrix}
1 & 0 & 0 & 0 & 0 & 0 & 0 & 0 & 0 \\
0 & 0 & 0 & 2 & 0 & 0 & 1 & 0 & 0 \\
2 & 0 & 0 & 2 & 0 & 0 & 2 & 0 & 0 \\
0 & 2 & 1 & 0 & 0 & 0 & 0 & 0 & 0 \\
0 & 0 & 0 & 0 & 2 & 1 & 0 & 1 & 0 \\
0 & 0 & 0 & 0 & 0 & 0 & 0 & 2 & 1 \\
2 & 2 & 2 & 0 & 0 & 0 & 0 & 0 & 0 \\
0 & 0 & 0 & 2 & 2 & 2 & 0 & 0 & 0 \\
0 & 0 & 0 & 0 & 0 & 0 & 2 & 2 & 2
\end{bmatrix}.
$$

If the Haar functions are written as rows of the Galois-Field-Haar matrix, then this matrix is

$$
\mathbf{GFH_3(2)}^{T} =
\begin{bmatrix}
1 & 1 & 1 & 1 & 1 & 1 & 1 & 1 & 1 \\
0 & 0 & 0 & 1 & 1 & 1 & 2 & 2 & 2 \\
0 & 0 & 0 & 1 & 1 & 1 & 1 & 1 & 1 \\
0 & 1 & 2 & 0 & 0 & 0 & 0 & 0 & 0 \\
0 & 0 & 0 & 0 & 1 & 2 & 0 & 0 & 0 \\
0 & 0 & 0 & 0 & 0 & 0 & 0 & 1 & 2 \\
0 & 1 & 1 & 0 & 0 & 0 & 0 & 0 & 0 \\
0 & 0 & 0 & 0 & 1 & 1 & 0 & 0 & 0 \\
0 & 0 & 0 & 0 & 0 & 0 & 0 & 1 & 1
\end{bmatrix}.
$$

If this matrix is used as the direct Galois-Field-Haar transform matrix, then the inverse transform is defined by the transposed matrix of $\mathbf{GFH_3(2)}^{-1}$. Thus,

$$
((\mathbf{GFH_3(2)})^{T})^{-1} =
\begin{bmatrix}
1 & 0 & 2 & 0 & 0 & 0 & 2 & 0 & 0 \\
0 & 0 & 0 & 2 & 0 & 0 & 2 & 0 & 0 \\
0 & 0 & 0 & 1 & 0 & 0 & 2 & 0 & 0 \\
0 & 2 & 2 & 0 & 0 & 0 & 0 & 2 & 0 \\
0 & 0 & 0 & 0 & 2 & 0 & 0 & 2 & 0 \\
0 & 0 & 0 & 0 & 1 & 0 & 0 & 2 & 0 \\
0 & 1 & 2 & 0 & 0 & 0 & 0 & 0 & 2 \\
0 & 0 & 0 & 0 & 0 & 2 & 0 & 0 & 2 \\
0 & 0 & 0 & 0 & 0 & 1 & 0 & 0 & 2
\end{bmatrix}.
$$

4.11.3 RECURRENCE RELATIONS FOR HAAR FUNCTIONS

The generalization of the Haar functions to $P(C_p^n)$ derived from decision diagrams is proper, since it involves the already defined Haar functions and extends the notion of Haar functions to different algebraic structures. Further, this generalization permits to derive recurrence relations for generation of Haar functions similar to that for the Haar transform in $C(C_2^n)$. These relations give an explicit relationship to the Fourier transforms and polynomial expressions in $P(C_p^n)$ and can be used as an alternative definition of the Haar functions.

Assume that for $f \in P(C_p^n)$ given by the vector of function values $\mathbf{F} = [f(0), \ldots, f(p^n - 1)]^T$, and that a series expansion for f is defined as

$$
\begin{aligned}
\mathbf{S}_f &= \mathbf{Q}^{-1}(n)\mathbf{F}, \\
\mathbf{F} &= \mathbf{Q}(n)\mathbf{S}_f,
\end{aligned}
$$

where $\mathbf{S}_f = [S_f(0), \ldots, S_f(p^n - 1)]^T$ is the vector of coefficients in the expansion with respect to the basis Q defined by columns of a matrix

$$
\mathbf{Q}(n) = \bigotimes_{i=1}^{n} \mathbf{Q}(1),
$$

where $\mathbf{Q}(1)$ is the Fourier transform matrix in $P(C_p)$ or the matrix defining the product terms in polynomial expressions in $P(C_p)$.

Definition 40 *Haar functions in $P(C_p^n)$ are defined by the relation*

$$
\mathbf{H}(n) = \left[
\begin{array}{c}
\mathbf{H}(n-1) \otimes \mathbf{q}_0 \\
\mathbf{I}(n-1) \otimes \mathbf{q}_1 \\
\vdots \\
\mathbf{I}(n-1) \otimes \mathbf{q}_{p-1}
\end{array}
\right],
$$

where $\mathbf{q}_i, i = 0, \ldots p - 1$ are rows of $\mathbf{Q}(1)^T$ for $P = GF(p)$, and of $\mathbf{Q}(1)^*$ for $P = C$.

Example 70 *Table 4.3 shows examples of recurrence relations for Haar functions for different choices of the domain groups C_p and ranges P and $n = 2$. For simplicity, the normalization constant are omitted. The identity matrices in $P(C_p)$ are denoted by $\mathbf{I}_p$.*

Definition 40 determines an algorithm for the design of Haar transforms in $P(C_p^n)$ for different choices of domain C_p and range P.

Algorithm 5 *(Design of Haar transforms)*

1. *Given a vector space $P(C_p^n)$.*

2. *Consider the matrix $\mathbf{Q}(1)$ defining the Fourier basis or the basis for polynomial expressions in $P(C_p)$ and define $\mathbf{H}(1) = \mathbf{Q}(1)^T$ if $P = GF(p)$ and $\mathbf{H}(1) = \mathbf{Q}(1)^*$ if $P = C$.*

3. *Denote the rows of* $\mathbf{H}(1)$ *as* $\mathbf{q}_0, \ldots, \mathbf{q}_{p-1}$.

4. *Determine the Haar matrix in* $P(C_p)$ *by performing the Kronecker product as is determined in Definition 40 and enumerate rows of the produced matrix as the Haar functions* $har(w, x)$, $w, x \in \{0, \ldots, p^n - 1\}$ *in* $P(C_p)$.

End of Algorithm.

The Haar transforms defined by using this algorithm produce the related HSTDDs. At the same time, they can be calculated through MTDDs by a simple generalization of DD methods for the Haar transform in $C(C_2^n)$. The generalization consists in application of the related matrices $\mathbf{H}(1)$ instead of $\mathbf{W}(1)$ and identity matrices $\mathbf{I}_p$ instead of $\mathbf{I}_2$. Thus, the FFT (or more properly FHT) algorithms can be derived.

Table 4.3: Recurrence relations for Haar functions

Domain	Range	Q	Haar transform
C_2	C	$\mathbf{W}(1)$	$\mathbf{H}(n) = \begin{bmatrix} \mathbf{W}(1) \otimes \begin{bmatrix} 1 & 1 \end{bmatrix} \\ \mathbf{I}_2 \otimes \begin{bmatrix} 1 & -1 \end{bmatrix} \end{bmatrix}$
C_2	$GF(2)$	$\mathbf{R}(1)$	$\mathbf{RMH}(n) = \begin{bmatrix} \mathbf{R}(1)^T \otimes \begin{bmatrix} 1 & 1 \end{bmatrix} \\ \mathbf{I}(1) \otimes \begin{bmatrix} 0 & 1 \end{bmatrix} \end{bmatrix}$
C_3	C	$\mathbf{VC}_3(1)$	$\mathbf{VCH}_3(n) = \begin{bmatrix} \mathbf{G}^*_{3GF}(1) \otimes \begin{bmatrix} 1 & 1 & 1 \end{bmatrix} \\ \mathbf{I}_3(1) \otimes \begin{bmatrix} 1 & e_2 & e_1 \end{bmatrix} \\ \mathbf{I}_3(1) \otimes \begin{bmatrix} 1 & e_1 & e_2 \end{bmatrix} \end{bmatrix}$
C_3	$GF(3)$	$\mathbf{G}_{3GF}(1)$	$\mathbf{GFH}_3(n) = \begin{bmatrix} \mathbf{G}^T_{3GF}(1) \otimes \begin{bmatrix} 1 & 1 & 1 \end{bmatrix} \\ \mathbf{I}_3(1) \otimes \begin{bmatrix} 0 & 1 & 2 \end{bmatrix} \\ \mathbf{I}_3(1) \otimes \begin{bmatrix} 0 & 1 & 1 \end{bmatrix} \end{bmatrix}$
C_4	C	$\mathbf{VC}_4(1)$	$\mathbf{VCH}_4(n) = \begin{bmatrix} \mathbf{VC}^*_3(1) \otimes \begin{bmatrix} 1 & 1 & 1 & 1 \end{bmatrix} \\ \mathbf{I}_4(1) \otimes \begin{bmatrix} 1 & i & -1 & -i \end{bmatrix} \\ \mathbf{I}_4(1) \otimes \begin{bmatrix} 1 & -1 & 1 & -1 \end{bmatrix} \\ \mathbf{I}_4(1) \otimes \begin{bmatrix} 1 & -i & -1 & i \end{bmatrix} \end{bmatrix}$
C_4	$GF(4)$	$\mathbf{G}_{4GF}(1)$	$\mathbf{GFH}_4(n) = \begin{bmatrix} \mathbf{G}^T_{4GF}(1) \otimes \begin{bmatrix} 1 & 1 & 1 & 1 \end{bmatrix} \\ \mathbf{I}_4(1) \otimes \begin{bmatrix} 0 & 1 & 2 & 3 \end{bmatrix} \\ \mathbf{I}_4(1) \otimes \begin{bmatrix} 0 & 1 & 3 & 2 \end{bmatrix} \\ \mathbf{I}_4(1) \otimes \begin{bmatrix} 0 & 1 & 1 & 1 \end{bmatrix} \end{bmatrix}$

CHAPTER 5

Fast Calculation Algorithms

Signal processing approach to logic functions allows to consider their functional expressions and spectral representations in a unified manner. Thus, it allows the usage of fast computing algorithms developed in signal processing to compute the coefficients in either functional expressions or spectral representations in the same way and by using algorithms of the identical structure by varying just the basic kernels of the algorithms and ranges were the computations are performed.

Since for the domain of an n-variable function $f(x_1, x_2, \ldots, x_n)$, we assume a group G of order g that is the direct product of subgroups G_i, $i = 1, 2, \ldots, n$, with $x_i \in G_i$, it follows that computing the coefficients in related representations for f can be expressed in matrix notation as

$$\mathbf{S}_f = \mathbf{Q} \cdot \mathbf{F},$$

where $\mathbf{F}$ is the function vector for f, $\mathbf{S}_f$ is the vector of coefficients that should be computed, and $\mathbf{Q}$ is a Kronecker product representable matrix defined in terms of the corresponding transform matrices $\mathbf{Q}_i$ on G_i. Due to this, we can use a theorem stating that

$$\mathbf{Q}_1 \otimes \mathbf{Q}_2 \otimes \cdot \otimes \mathbf{Q}_n = \mathbf{C}_1 \cdot \mathbf{C}_2 \cdot \ldots \cdot \mathbf{C}_n,$$

where $(g \times g)$ matrices are again Kronecker product representable as

$$\mathbf{C}_i = \otimes_{k=1}^{n} \mathbf{R}_k,$$

with

$$\mathbf{R}_k = \begin{cases} \mathbf{Q}_i, & k = i, \\ \mathbf{I}_{g_i}, & \text{otherwise}, \end{cases}$$

where $\mathbf{I}_{g_i}$ is the $(g_i \times g_i)$ identity matrix.

Since the Kronecker product for $\mathbf{C}_i$ involves the identity matrices $\mathbf{I}_{g_i}$, it follows that $\mathbf{C}_i$ are sparse matrices leading to a fast computation algorithm.

In signal processing such factorization for $\mathbf{Q}$ is called the Good-Thomas factorization, and the corresponding fast algorithm belongs to the Good-Thomas FFT-like algorithms; for instance, see [195].

The computations specified by the component matrices $\mathbf{Q}_i$ are called the basic butterfly operations and represent kernels of the fast computation algorithm. The algorithm consists of n steps, each step described by a matrix $\mathbf{C}_i$, $i = 1, 2, \ldots, n$. Within each step, the computations are performed by implementing the same kernel (butterfly) corresponding to $\mathbf{Q}_i$ over different and disjoint sets of data, which provides high parallelization possibilities in this essentially sequential algorithm since the input data of a step are the output data from the previous step. The intermediate results are independent of each other, which enables in-place computations in the sense that results of computation in a step are saved in the same memory locations instead of the results of computations in the previous step. The algorithm requires the minimum number of operations to perform the computations of the form $\mathbf{Q} \cdot \mathbf{F}$, and the complexity of the algorithm is $O(g \log g)$, where g is the order of G, i.e., the number of points where f is defined.

When the algorithm is expressed in matrix notation, then function vectors to represent the input function to be processed, the intermediate computations, and required coefficients at the output are a natural choice. The problem can be the size g of vectors, which is exponential in the number of variables in f. Therefore, the same algorithms can be equally implemented over other data structures to represent logic functions, such as cubes, arrays, and decision diagrams. These algorithms are also used as kernels for various other algorithms for different tasks in theory and practical applications of logic functions. For other approaches to compute the coefficients of functional expressions and spectral representations of logic functions; for instance, see [233], [234], [235], [239], [240], [241].

5.1   ILLUSTRATIVE EXAMPLES OF FFT-LIKE ALGORITHMS

In this section, we will illustrate the implementation of FFT-like algorithms over function vectors and decision diagrams on the example of ternary functions of three variables.

$$\mathbf{GF}_{3GF} = \begin{bmatrix} 1 & 0 & 0 \\ 0 & 2 & 1 \\ 2 & 2 & 2 \end{bmatrix}$$

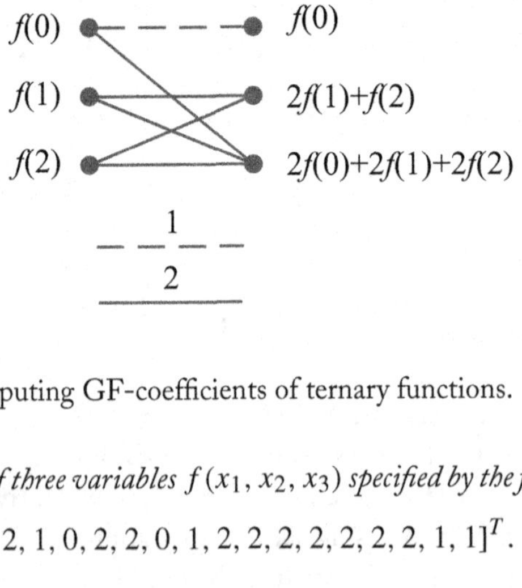

Figure 5.1: The butterfly operation for computing GF-coefficients of ternary functions.

Example 71 *Consider the ternary function of three variables $f(x_1, x_2, x_3)$ specified by the function vector*

$$\mathbf{F} = [1, 1, 1, 1, 1, 1, 1, 1, 1, 0, 2, 1, 0, 2, 2, 0, 1, 2, 2, 2, 2, 2, 2, 2, 1, 1]^T.$$

The GF-transform matrix is

$$\begin{aligned} \mathbf{G}_{3GF}(3) &= \mathbf{G}_{3GF}(1) \otimes \mathbf{G}_{3GF}(1) \otimes \mathbf{G}_{3GF}(1) \\ &= \begin{bmatrix} 1 & 0 & 0 \\ 0 & 2 & 1 \\ 2 & 2 & 2 \end{bmatrix} \otimes \begin{bmatrix} 1 & 0 & 0 \\ 0 & 2 & 1 \\ 2 & 2 & 2 \end{bmatrix} \otimes \begin{bmatrix} 1 & 0 & 0 \\ 0 & 2 & 1 \\ 2 & 2 & 2 \end{bmatrix}. \end{aligned}$$

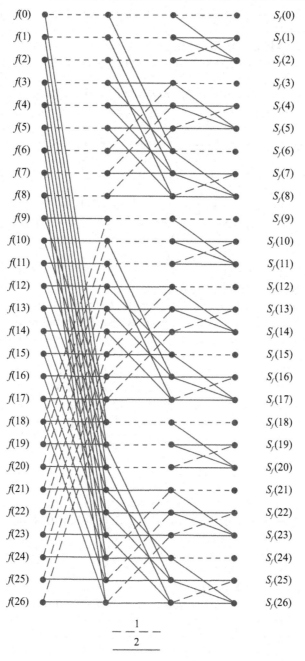

Figure 5.2: Flow-graph of the FFT-like algorithm for computing the coefficients of GF-expressions of ternary functions of three variables.

By using this matrix, we compute the coefficients in the GF-expression for f as

$$
\begin{aligned}
\mathbf{S}_{f,3GF} &= \mathbf{G}_{3GF}(3)\mathbf{F} \\
&= [1, 0, 0, 0, 0, 0, 0, 0, 0, 1, 1, 0, 2, 1, 2, 1, 2, 1, 2, 1, 0, 2, 1, 1, 1, 2, 2]^T.
\end{aligned}
$$

The same computation can be performed by the FFT-like algorithm based upon the factorization of the GF-transform matrix as

$$
\mathbf{G}_{3GF}(3) = \mathbf{C}_1 \cdot \mathbf{C}_2 \cdot \mathbf{C}_3,
$$

where

$$
\begin{aligned}
\mathbf{C}_1 &= \mathbf{G}_{3GF}(1) \otimes \mathbf{I}_3(1) \otimes \mathbf{I}_3(1), \\
\mathbf{C}_2 &= \mathbf{I}_3(1) \otimes \mathbf{G}_{3GF}(1) \otimes \mathbf{I}_3(1), \\
\mathbf{C}_3 &= \mathbf{I}_3(1) \otimes \mathbf{I}_3(1) \otimes \mathbf{G}_{3GF}(1).
\end{aligned}
$$

Each of these matrices $\mathbf{C}_i$, $i = 1, 2, 3$, describes a step in the FFT-like algorithm for computation of the coefficients in the GF-expressions for ternary functions of three variables. The basic transform matrix $\mathbf{G}_{3GF}(1)$ specifies the butterfly operation in this algorithm as shown in Fig. 5.1.

Fig. 5.2 shows the flow graph of the algorithm and the computations are performed as follows:

$$
\begin{aligned}
\mathbf{S}_{f,3GF} &= \mathbf{G}_{3GF}(3)\mathbf{F} \\
&= \mathbf{C}_1\mathbf{C}_2\mathbf{C}_3\mathbf{F} \\
&= \mathbf{C}_1\mathbf{C}_2(\mathbf{C}_3\mathbf{F}) \\
&= \mathbf{C}_1(\mathbf{C}_2\mathbf{C}_3\mathbf{F}) \\
&= \mathbf{C}_1\mathbf{C}_2\mathbf{C}_3\mathbf{F}.
\end{aligned}
$$

1. Step 1

$$
\begin{aligned}
\mathbf{S}_{f_1,3GF} &= \mathbf{C}_1\mathbf{F} \\
&= [1, 1, 1, 1, 1, 1, 1, 1, 1, 2, 0, 1, 2, 0, 0, 1, 0, 2, 0, 1, 2, 0, 1, 1, 1, 0]^T.
\end{aligned}
$$

2. Step 2

$$
\begin{aligned}
\mathbf{S}_{f_{1,2},3GF} &= \mathbf{C}_1\mathbf{S}_{f_1,3GF} \\
&= [1, 1, 1, 0, 0, 0, 0, 0, 0, 1, 2, 0, 2, 2, 0, 1, 1, 0, 2, 0, 1, 2, 1, 2, 1, 2, 1]^T.
\end{aligned}
$$

3. Step 3

$$
\begin{aligned}
\mathbf{S}_{f_{1,2,3},3GF} &= \mathbf{C}_1\mathbf{S}_{f_{1,2},3GF} \\
&= [1, 0, 0, 0, 0, 0, 0, 0, 0, 1, 1, 0, 2, 1, 2, 1, 2, 1, 2, 1, 0, 2, 1, 1, 1, 2, 2]^T. \\
&= \mathbf{S}_{f,3GF}.
\end{aligned}
$$

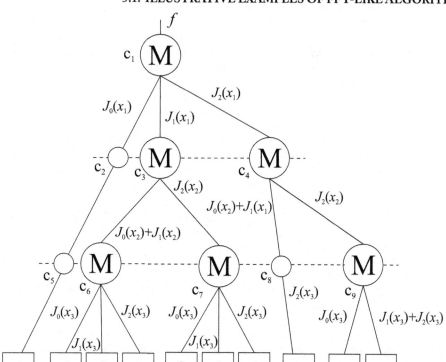

Figure 5.3: MDD for the function f in Example 72.

The same computations as in the above example can be performed assuming the decision diagram representations as the underlying data structure.

Example 72 *Figure 5.3 shows the MDD for the considered function f. In this figure, the symbol $a + b$ means that both edges labeled by a and b point to the same node.*

The steps of the FFT-like algorithm described above can be performed over this MDD by processing each node in the diagram by the basic butterfly operation specified by $\mathbf{G}_{3GF}(1)$. The processing means that the inputs to the butterfly operation are subfunctions represented by the subdiagrams rooted at the nodes pointed by the outgoing edges of the processed node. For the clarity of presentation, we will express the impact of deleted nodes through cross points shown by small circles in Fig. 5.3 and viewed as crossings of a path in the diagram with an imaginary line connecting nodes at the same level in the diagram. In practical programming implementations, these computations are avoided and the procedure simplified by using properties of the performed transforms. In particular, the computations are reduced to transforming the related subfunctions and padding with zeros. It can be followed, depending on the transform, by the multiplication of the constant value of a terminal node or the subfunction pointed by the edge of the processed

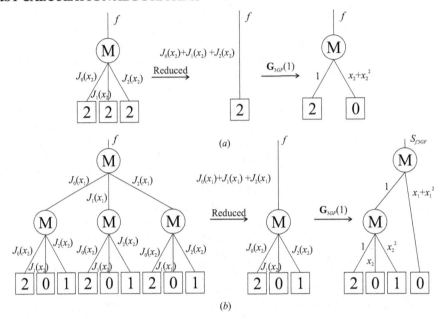

Figure 5.4: Illustration of computing the impact of deleted nodes in MDD.

node by a constant value equal to the length of the path between these two nodes. The explanation for this implementation is the following. If a node is deleted from the MDD, this means that the outgoing edges of this node point to the identical subfunctions. Therefore, in the considered case of ternary functions, nodes have three outgoing edges and the subfunction represented by the deleted node has three identical parts. Thus, this node represents a periodic subfunction or a constant. Then, due to the properties of the transforms, the spectrum of a constant function is the delta function, while the spectrum of a periodic function is the delta function of the length of a period Kronecker multiplied by the spectrum of the periodically repeated subfunction in the considered periodic functions.

Figure 5.4 illustrates computing the impact of the deleted nodes to the GF-spectrum for ternary functions for the case of nodes whose outgoing edges point to the identical constant values and the identical subfunctions. For the node in Fig. 5.4(a), the computation is obviously done as follows:

$$
\mathbf{S}_f = \mathbf{G}_{3GF}(3)
\begin{bmatrix} 2 \\ 2 \\ 2 \end{bmatrix}
=
\begin{bmatrix} 1 & 0 & 0 \\ 0 & 2 & 1 \\ 2 & 2 & 2 \end{bmatrix}
\begin{bmatrix} 2 \\ 2 \\ 2 \end{bmatrix}
=
\begin{bmatrix} 2 \\ 0 \\ 0 \end{bmatrix}.
$$

The computations for the node in Fig. 5.4(a) are done as

$$
\mathbf{S}_f = \mathbf{G}_{3GF}(3)
\begin{bmatrix}
\begin{bmatrix} 2 \\ 0 \\ 1 \end{bmatrix} \\
\begin{bmatrix} 2 \\ 0 \\ 1 \end{bmatrix} \\
\begin{bmatrix} 2 \\ 0 \\ 1 \end{bmatrix}
\end{bmatrix}
=
\begin{bmatrix}
1 \cdot \begin{bmatrix} 2 \\ 0 \\ 1 \end{bmatrix} \\
2 \cdot \begin{bmatrix} 2 \\ 0 \\ 1 \end{bmatrix} \oplus 1 \cdot \begin{bmatrix} 2 \\ 0 \\ 1 \end{bmatrix} \\
2 \cdot \begin{bmatrix} 2 \\ 0 \\ 1 \end{bmatrix} \oplus 2 \cdot \begin{bmatrix} 2 \\ 0 \\ 1 \end{bmatrix} \oplus 2 \cdot \begin{bmatrix} 2 \\ 0 \\ 1 \end{bmatrix}
\end{bmatrix}
=
\begin{bmatrix}
2 \\ 0 \\ 1 \\ 0 \\ 0 \\ 0 \\ 0 \\ 0 \\ 0
\end{bmatrix}.
$$

If the nodes and cross points are labeled as in Fig. 5.3, the GF-coefficients for the considered function
f are computed as follows.

Step 1

$$
\mathbf{c}_5 = \mathbf{G}_{3GF}(1) \begin{bmatrix} 1 \\ 1 \\ 1 \end{bmatrix} = \begin{bmatrix} 1 \\ 0 \\ 0 \end{bmatrix}
\qquad
\mathbf{c}_6 = \mathbf{G}_{3GF}(1) \begin{bmatrix} 1 \\ 0 \\ 2 \end{bmatrix} = \begin{bmatrix} 1 \\ 2 \\ 0 \end{bmatrix}
$$

$$
\mathbf{c}_7 = \mathbf{G}_{3GF}(1) \begin{bmatrix} 2 \\ 0 \\ 1 \end{bmatrix} = \begin{bmatrix} 2 \\ 1 \\ 0 \end{bmatrix}
\qquad
\mathbf{c}_8 = \mathbf{G}_{3GF}(1) \begin{bmatrix} 2 \\ 2 \\ 2 \end{bmatrix} = \begin{bmatrix} 2 \\ 0 \\ 0 \end{bmatrix}
$$

$$
\mathbf{c}_9 = \mathbf{G}_{3GF}(1) \begin{bmatrix} 2 \\ 1 \\ 1 \end{bmatrix} = \begin{bmatrix} 2 \\ 0 \\ 2 \end{bmatrix}.
$$

Step 2

$$
\mathbf{c}_2 = \mathbf{G}_{3GF}(1)
\begin{bmatrix} \mathbf{c}_5 \\ \mathbf{c}_5 \\ \mathbf{c}_5 \end{bmatrix}
=
\begin{bmatrix}
\begin{bmatrix} 1 \\ 0 \\ 0 \end{bmatrix} \\
\begin{bmatrix} 1 \\ 0 \\ 0 \end{bmatrix} \\
\begin{bmatrix} 1 \\ 0 \\ 0 \end{bmatrix}
\end{bmatrix}
=
\begin{bmatrix}
1 \\ 0 \\ 0 \\ 0 \\ 0 \\ 0 \\ 0 \\ 0 \\ 0
\end{bmatrix}
$$

$$\mathbf{c}_3 = \mathbf{G}_{3GF}(1) \begin{bmatrix} \mathbf{c}_6 \\ \mathbf{c}_6 \\ \mathbf{c}_7 \end{bmatrix} = \begin{bmatrix} \begin{bmatrix} 1 \\ 2 \\ 0 \end{bmatrix} \\ \begin{bmatrix} 1 \\ 2 \\ 0 \end{bmatrix} \\ \begin{bmatrix} 2 \\ 1 \\ 0 \end{bmatrix} \end{bmatrix} = \begin{bmatrix} 1 \\ 2 \\ 0 \\ 1 \\ 2 \\ 0 \\ 2 \\ 1 \\ 0 \end{bmatrix}$$

$$\mathbf{c}_4 = \mathbf{G}_{3GF}(1) \begin{bmatrix} \mathbf{c}_8 \\ \mathbf{c}_8 \\ \mathbf{c}_9 \end{bmatrix} = \begin{bmatrix} \begin{bmatrix} 2 \\ 0 \\ 0 \end{bmatrix} \\ \begin{bmatrix} 2 \\ 0 \\ 0 \end{bmatrix} \\ \begin{bmatrix} 2 \\ 0 \\ 2 \end{bmatrix} \end{bmatrix} = \begin{bmatrix} 2 \\ 0 \\ 0 \\ 0 \\ 0 \\ 2 \\ 0 \\ 0 \\ 1 \end{bmatrix}.$$

Step 3

$$\mathbf{c}_1 = \mathbf{G}_{3GF}(1) \begin{bmatrix} \mathbf{c}_2 \\ \mathbf{c}_3 \\ \mathbf{c}_4 \end{bmatrix}$$

$$= \begin{bmatrix} [1,0,0,0,0,0,0,0,0]^T & [1,2,0,1,2,0,2,1,0]^T & [2,0,0,0,0,2,0,0,1]^T \end{bmatrix}^T$$

$$= [1,0,0,0,0,0,0,0,0,1,1,0,2,1,2,1,2,1,2,1,0,2,1,1,1,2,2]^T.$$

Each step of the computation can be represented by MDD, which then can be combined into the MDD for the GF-coefficients. From spectral interpretation of decision diagrams, this MDD for the GF-spectrum of f after conversion of meaning of nodes and correspondingly labels at the edges that becomes the GFDD for f. Figure 5.5 shows this GFDD for the considered function f. It can be noticed that, counting from the left to the right, the third and fourth node for x_3, as well as the sixth and seventh node for the same variable, represent, respectively, subfunctions that are identical to each other multiplied by 2. Thus, the diagram can be simplified by allowing edges with multiplicative attributes in the same way as this is done in BDDs with negated edges; see [170].

The same methods illustrated here for the case of GF-coefficients can be equally applied to compute the coefficients of other Kronecker or layer-Kronecker product representable transform-matrices [221], [222]. It follows that they are also used to compute the Haar and Haar-like transforms,

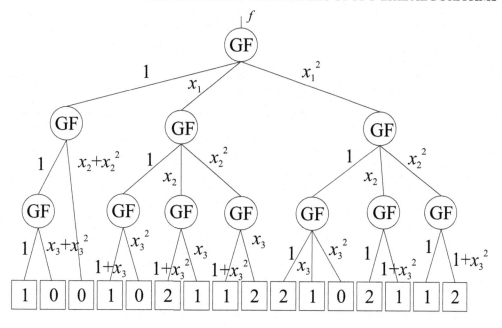

Figure 5.5: GFDD for f in Example 72.

for example, see [99]. In the case of Haar and related transform, the algorithms over decision diagrams can be implemented for an order of magnitude faster, by using properties of the Haar and Haar related functions in terms of which these transforms are defined. For more details for computing the Haar transform for functions of binary variables, see [178], [179], [180], [181], [219], since the same methods extends to multiple-valued functions straightforwardly.

Bibliography

[1] Adams, K., *Optimization of Multiple-Valued Polynomials by Polarities and Affine Transforms*, PhD Thesis, University of Ulster, Northern Ireland, UK, March 2007. Cited on page(s) 40

[2] Adams, K.J., McGregor, J., "Comparison of different features of quaternary Reed-Muller canonical forms and some new statistical results," *Proc. 32nd Int. Symp. on Multiple-Valued Logic*, Boston, Massachusetts, USA, May 2002, 83-88. DOI: 10.1109/ISMVL.2002.1011074 Cited on page(s) 40

[3] Adams, K.J., McGregor, J., "New information on the effectiveness of different Reed-Muller algebras on the representation of quaternary functions," *Proc. 33rd Int. Symp. on Multiple-Valued Logic*, Tokyo, Japan, May 16-19, 2003, 33-39. DOI: 10.1109/ISMVL.2003.1201381 Cited on page(s) 40

[4] Adams, K.J., McGregor, J., "On the optimisation of Reed-Muller expressions," *Proc. 34th International Symposium on Multiple-Valued Logic*, Toronto, Canada, May 19-22, 2004, 168-176. DOI: 10.1109/ISMVL.2004.1319937 Cited on page(s) 40

[5] Aizenberg, I.N., Aizenberg, N.N., Vandewalle J. *Multi-Valued and Universal Binary Neurons - Theory, Learning, Applications*, Kluwer Academic Publishers, Boston, Dordrecht, London, 2000. Cited on page(s) 8

[6] Aizenberg, N.N., Kukharev, G.A., Pak, I.O., Shmerko, V.D., "Methods of solution of logical equations by using fast orthogonal transforms and techniques of their realizations" in *IX Vsesoyuznoe sovetschanie po problemam upravleniya*, Tezisy doklada, Moscow, 1983. Cited on page(s) 8

[7] Aizenberg, N.N., Rabinovich, Z.L., "Some classes of functional systems of operations and canonical forms of many-valued logical functions," *Kibernetika*, No. 2, 1965, 37-45, in Russian. Cited on page(s) 8

[8] Aizenberg, N.N., Trofimljuk, O.T., "Conjunctive transforms for discrete signals and their applications of tests and the detection of monotone functions," *Kibernetika*, No. 5, K, 1981, in Russian. Cited on page(s) 8

[9] Allen, C.M., Givone, D.D., "A minimization technique formultiple-valued logic systems," *IEEE Trans. Comput.*, Vol. 17, 1968, 182–184. DOI: 10.1109/TC.1968.227407 Cited on page(s) 11

128   BIBLIOGRAPHY

[10] Antonenko, V., Guvakov, I., Shmerko, V., Kaczmarek, A., Yanushkevich, S., "Linear arithmetical forms of k-valued functions," *Proc. European Conf. on Circuit Theory and Design*, Turkey, 1995, 323-328. Cited on page(s) 48

[11] Astola, J.T., Stanković, R.S., *Fundamentals of Switching Theory and Logic Design*, Springer, 2006. Cited on page(s) 11, 16

[12] Astola, J., Stanković, R.S., "Signal processing algorithms and multiple-valued logic design methods," *Proc. 36th Int. Symp. on Multiple-Valued Logic*, May 17-20, 2006, Singapore, 16/1 - 16/8. DOI: 10.1109/ISMVL.2006.38 Cited on page(s) 11

[13] Astola, J.T., Stanković, R.S., "Application of Covering Codes in Determination of Sparse Representations of Switching Functions," *Proc. 39th Int. Symp. on Multiple-Valued Logic*, Naha, Okinawa, Japan, May 21-23, 1009, 304-311. Cited on page(s) 54

[14] Benjauthrit, B., Reed, I.S., "Galois switching functions and their applications," *IEEE Trans. Computers*, Vol.C-25, No.1, 1976, 79-86. Cited on page(s) 8, 25

[15] Berlin, R.D., "Synthesis of N-valued switching circuits," *IRE Trans. Electron. Comput*, Vol. EC-7, 1958, 52-56. DOI: 10.1109/TEC.1958.5222096 Cited on page(s) 11

[16] Bernstein, B.A., "Modular representations of finite algebras," *Proc. 7th Int. Congress Mathematicians*, Univ. Toronto Press, 1928, Vol. 1, 1924, 207-216. Cited on page(s) 11

[17] Besslich, Ph.W., "Efficient computer method for EXOR logic design," *IEE Proc.*, Vol. 131, Pt.E, No. 6, 1983, 203-206. Cited on page(s) 16

[18] Boole, G., *The Matehamtical Analysis of Logic - Being an Essay Towards a Calculus of Deductive Reasoning*, Publisher MacMilland, Barclay, & MacMillan, Cambridge, George Bell, London, 1847. Cited on page(s) 1

[19] Boole, G., *An Investigation of the Laws of Thought on which are Founded the Mathematical Theories of Logic and Probabilities*, Publisher Walton and Maberly, London, MacMillan and Co., Cambridge, 1854. Cited on page(s) 1

[20] Bryant, R.E., "Graph-based algorithms for Boolean functions manipulation," *IEEE Trans. Computers*, Vol.C-35, No.8, 1986, 667-691. Cited on page(s) 78

[21] Bryant, R.E., Chen, Y.A., "Verification of arithmetic functions with binary moment decision diagrams," unpublished paper, May 31, 1994, CMU-CS-94-160. Cited on page(s) 100, 106, 107, 108

[22] Brayton, R.K., "The future of logic synthesis and verification," in Hassoun, S., Sasao, T., (eds.), *Logic Synthesis and Verification*, Kluwer Academic Publishers, Boston, MA, USA, 2002, 403-434. Cited on page(s) 8

[23] Butler, J.T., Sasao, T., "On the properties of multiple-valued functions that are symmetric in both variable values and labels," *Proc. 28th Int. Symp. on Multiple-Valued Logic*, Fukuoka, Japan, May 27-29, 1998, 83-88. DOI: 10.1109/ISMVL.1998.679299 Cited on page(s) 12

[24] Butler, J.T., Schueller, K.A., "Worst case number of terms in symmetric multiple-valued functions," *Proc. 21st Inx. Symp. on Multiple-Valued Logic*, Victoria, BC, Canada, May 26-29, 1991, 94-101. DOI: 10.1109/ISMVL.1991.130712 Cited on page(s) 12

[25] Chen, X., Moraga, C., "An algebra for current-mode CMOS multivalued circuits," *Proc. 23rd Int. Symp. on Multi-Valued Logic*, 1993, 239-244. DOI: 10.1109/ISMVL.1993.289553 Cited on page(s) 11

[26] Chen, X., Moraga, C., "Design of multivalued circuits based on an algebra for current-Mode CMOS multivalued circuits," *J. of Comput. Sci. & Technol.*, Vol. 10, No. 6, 1995, 565-568. DOI: 10.1007/BF02943514 Cited on page(s) 11

[27] Lee, C.U., Han, L. S.-I., Kim, J.-O., Kim, H.-S.,"The generation sequential method to generalized Reed-Muller coefficients over GF(3)," *Proc. of the IEEE Region 10 Conference*, (TENCON 99), Vol. 1, 1999, 387-390. Cited on page(s) 49

[28] Chrzanowska-Jeske, M., Xu, Y., Perkowski, M.A., "Logic synthesis for a regular layout," *VLSI Design*, Vol. 12, No. 3, 2000. DOI: 10.1155/1999/85272 Cited on page(s) 12

[29] Cignoli, R., "The algebras of Lukasiewicz many-valued logic - A historical overview," in S. Aguzzoli et al.(Eds.), *Algebraic and Proof-theoretic Aspects*, LNAI 4460, Springer, 2007, 69-83. Cited on page(s) 9

[30] Clarke, E.M., McMillan, Zhao, X., Fujita, M., "Spectral transforms for extremely large Boolean functions," in: Kebschull, U., Schubert, E., Rosentiel, W., Eds., *Proc. IFIP WG 10.5 Workshop on Applications of the Reed-Muller Expansion in Circuit Design*, 16-17.9.1993, Hamburg, Germany, 86-90. Cited on page(s) 16

[31] Cohn, M., *Switching Function Canonical Form over Integer Fields*, PhD thesis, Harvard University, Cambridge, Massachusetts, USA, December 1960. Cited on page(s) 8, 11

[32] Copeland, A.H., "Expansion of certain logical functions," *Amer. Math. Month.*, Vol. 44, No. 4, 1937, 213-218. DOI: 10.2307/2300689 Cited on page(s) 41

[33] Damarla, T., "Generalized Transforms for multiple-valued logic circuits and their fault detection," *IEEE Trans. Computers*, Vol. C-41, No. 9, 1992, 1101-1109. DOI: 10.1109/12.165392 Cited on page(s) 49

[34] De Morgan, A., *Formal Logic - or Calculus of Inference, Necessary and Probable*, London, 1874, xvi+336. Cited on page(s) 8

[35] Drechsler, R., Becker, B., Ruppertz, S., "The K*BMD - A verification data structure," *IEEE Design & Test Journal*, Vol. 14, No. 2, 1997, 51-59. DOI: 10.1109/54.587742 Cited on page(s) 100

[36] Drechsler, R., Janković, D., Stanković, R.S., "Generic implementation of DD packages in MVL," *Proc. 25th EUROMICRO Conference*, Vol. 1, 1999, 352-359.
DOI: 10.1109/EURMIC.1999.794491 Cited on page(s) 106

[37] Drechsler, R., Janković, D., Stanković, R.S., "Generic implementation of multi-valued logic decision diagram packages," *Multiple-Valued Logic and Soft Computing*, Vol. 11, No. 1-2, 2005. Cited on page(s) 106

[38] Dubrova, E., "Evaluation of m-valued fixed polarity generalizations of Reed-Muller canonical form," *Proc. 29th Int. Symp. on Multiple-Valued Logic*, May 20-22, 1999, 92-98. Cited on page(s) 27

[39] Dubrova, E.V., Muzio, J.C., "Testability of generalized multiple-valued Reed-Muller circuits," *Proc. 26th Int. Symp. on Multiple-Valued Logic*, 1996, 56-61.
DOI: 10.1109/ISMVL.1996.508336 Cited on page(s) 49

[40] Dubrova, E.V., Muzio, J.C., "Generalized Reed-Muller canonical form for a multiple-valued algebra," *Multi-Valued Logic Jr.*, 1996, Vol. 1, 65-84. Cited on page(s) 8, 11, 49

[41] Epstein, G., "Synthesis of electronic circuits for symmetric functions," *IRE Transactions on Electronic Computers*, Vol. EC-7, 1958, 57-59. DOI: 10.1109/TEC.1958.5222097 Cited on page(s) 11

[42] Epstein, G., "The lattice theory of Post algebras," *Trans. Amer. Math. Soc.*, Vol. 95, No. 2, 1960, 300-317. DOI: 10.1090/S0002-9947-1960-0112855-8 Cited on page(s) 11

[43] Epstein, G., "General synthesis of electronic circuits for symmetric functions," Computer Science Conference Abstracts, Columbus Ohio, 35, Feb. 1973. Cited on page(s) 11

[44] Epstein, G., "An equational axiomatization for the disjoint systems of Post algebras," *IEEE Trans. Computers*, Vol. 22, No. 4, 1973, 422-423. DOI: 10.1109/T-C.1973.223731 Cited on page(s) 11

[45] Epstein, G., *Multiple-Valued Logic Design: An Introduction*, IOP Publishing Ltd, 1993. Cited on page(s) 11

[46] Epstein, G., D.M. Miller and J.C. Muzio, "Some preliminary views on the general synthesis of electronic circuits for symmetric and partially symmetric functions," *Proc. 7th Int. Symp. on Multiple-Valued Logic*, May 1977, 29-34. Cited on page(s) 12

[47] Epstein, G., D.M. Miller and J.C. Muzio, "Selecting don't-care sets for many-valued functions: a pictorial approach using matrices," *Proc. 10th Int. Symp. on Multiple-Valued Logic*, June 1980, 219-225. Cited on page(s) 12

[48] Farm, P. Dubrova, E., Stanković, R.S., Astola, J., "Conjunctive decomposition for multiple-valued input binary-valued output functions," *Proc. TISCP Workshop on Spectral Methods and Multirate Signal Processing, SMMSP'02*, Toulouse, France, September 7-8, 2002, 227-234. Cited on page(s) 49

[49] Falkowski, B.J., "A note on the polynomial form of Boolean functions and related topics," *IEEE Trans. on Computers*, Vol. 48, No. 8, 1999, 860-864. DOI: 10.1109/12.795128 Cited on page(s) 41

[50] Falkowski, B.J., "Haar transform, calculation, generalization and applications in logic design," *Proc. 2nd Int. Workshop on Transforms and Filter Banks*, Brandenburg, Germany, March 1999, Edited by R. Creutzburg and J. Astola, TICSP Series # 4, March 2000, 101-120. Cited on page(s) 49

[51] Falkowski, B.J., Chang, C.H., "Efficient algorithm for the calculation of arithmetic spectrum from OBDD and synthesis of OBDD from arithmetic spectrum for incompletely specified Boolean functions," *Proc. IEEE Int. Symp. on Circuits and Systems ISCAS94*, USA, 1994. DOI: 10.1109/ISCAS.1994.408789 Cited on page(s) 49

[52] Falkowski, B.J., Chang, C.H., "Forward and inverse transformations between Haar spectra and ordered binary decision diagrams of Boolean functions," *IEEE Trans. on Computers*, Vol. 46, No. 11, 1997, 1272-1279. DOI: 10.1109/12.644301 Cited on page(s) 49

[53] Falkowski, B.J., Chang, C.-H., "Generalised k-variable-mixed-polarity Reed-Muller expansions for system of Boolean functions and their minimisation," *IEE Proc. Circuits, Devices and Systems*, Vol. 147, No. 4, 2000, 201-210. DOI: 10.1049/ip-cds:20000588 Cited on page(s) 27

[54] Falkowski, B.J., Fu, C., "Fastest classes of linearly independent transforms over $GF(3)$ and their properties," *IEE Proc. Computers and Digital Techniques*, Vol. 152, No. 5, 2005, 567-576. DOI: 10.1049/ip-cdt:20045162 Cited on page(s) 49, 54

[55] Falkowski, B.J., Holowinski, G., Malecki, K., "Effective minimization of logic functions in Reed-Muller domain," *Proc. Int. Conf. on Applications of Computer Systems*, Poland, 1997, 248-255. Cited on page(s) 49

[56] Falkowski, B.J., Lozano, C.C., "Quaternary Fixed-Polarity Reed-Muller expansion computation through operations on disjoint cubes and its comparison with other methods," *Computers and Electrical Engineering*, Vol. 31, No. 2, 2005, 112-131. DOI: 10.1016/j.compeleceng.2005.01.002 Cited on page(s) 49

[57] Falkowski, B.J., Lozano, C.C., Rahardja, S., "Spectra generation for fixed-polarity Reed-Muller transform over $GF(5)$," *Proc. 34th Int. Symp. on Multiple-Valued Logic*, Toronto, Canada, May 19-22, 2004, 177-183. DOI: 10.1109/ISMVL.2004.1319938 Cited on page(s) 49

[58] Falkowski, B.J., Rahardja, S., "Fast construction of polarity coefficient matrices for fixed polarity quaternary Reed-Muller expansions," *Proc. 5th Int. Workshop on Spectral Techniques*, 15.-17.3.1994, Beijing, China, 220-225. DOI: 10.1016/j.compeleceng.2005.01.002 Cited on page(s) 49

[59] Falkowski, B.J., Rahardja, S., "Efficient algorithm for the generation of fixed polarity quaternary Reed-Muller expansions," *Proc. 25th International Symposium on Multiple-Valued Logic*, 1995, 158-163. DOI: 10.1109/ISMVL.1995.513525 Cited on page(s) 18, 49, 54

[60] Falkowski, B.J., Rahardja, S., "Efficient computation of quaternary fixed polarity Reed-Muller expansions," *IEE Proc. Computers and Digital Techniques*, Vol. 142 No. 5, 1995, 345-352. DOI: 10.1049/ip-cdt:19952126 Cited on page(s) 18, 49

[61] Falkowski, B.J., Rahardja, S., "Novel quantized transform for ternary systems," *Proc. 25th Int. Symp. on Multiple-Valued Logic*, 23.-25.5.1995, Bloomington, Indiana, USA, 117-122. DOI: 10.1109/ISMVL.1995.513519 Cited on page(s) 49

[62] Falkowski, B.J., Rahardja, S., "Efficient computation of quaternary fixed polarity Reed-Muller expansions," *IEE Proc. Computers and Digital Techniques*, Vol. 142, No. 5, 1995, 345-352. DOI: 10.1049/ip-cdt:19952126 Cited on page(s) 27, 49

[63] Falkowski, B.J., Rahardja, S., "Generalised hybrid arithmetic canonical expansions for completely specified quaternary functions," *IEE Proc. Circuits, Devices and Systems*, Vol. 144, No. 4, 1997, 201-208. DOI: 10.1049/ip-cds:19970874 Cited on page(s) 49, 54

[64] Falkowski B., Shmerko V., Yanushkevich, S., "Arithmetic logic - its status and achievement," *Proc. Int. Conf on Applications of Computer Systems*, Szczecin, Poland, 1997, 208-223. Cited on page(s) 49

[65] Fu, C., Falkowski, B.J., "Linearly independent ternary Arithmetic Helix Transforms, Their Properties and Relations," *IEE Proc., Vision, Image and Signal Processing*, United Kingdom, Vol. 153, No. 2, 2006, 87-94. DOI: 10.1049/ip-vis:20045248 Cited on page(s) 8, 49, 54

[66] Garaev, M.U., Faradzhev, R.G., "On an analog of Fourier expressions over Galois fields and its applications to problems of generalized sequential machines," *Izv. Akad. Nauk Aizerb. SSR*, Ser. Fiz.-Techn, i Mat. Nauk, No. 6, 1968, 69-75. Cited on page(s) 8, 16

[67] Gibbs, E.J., "Harmonic analysis in the dyadic field regarded as a function space," Seminar, Royal Signals and Radar Estab., 28.10.1976, ii+24. Cited on page(s) 16, 101

[68] Gibbs, J.E., "Instant Fourier transform," *Electronics Letters*, Vol. 13, No.5, 1977, 122-123. DOI: 10.1049/el:19770086 Cited on page(s) 16, 28, 29, 44

[69] Gibbs, J.E., "Local and global views of differentiation," in Butzer, P.L., Stanković, R.S., (Eds.), *Theory and Applications of Gibbs Derivatives*, Matematički institut, Beograd, 1990, 1-19. Cited on page(s) 16, 101

[70] Green, D.H., "Reed-Muller expansions with fixed and mixed polarities over GF(4)," *IEE Proc. Computers and Digital Techniques*, Vol. 137, No. 5, 1990, 380-388. DOI: 10.1049/ip-e.1990.0047 Cited on page(s) 49

[71] Green, D.H., Taylor, I.S., "Modular representation of multiple-valued logic systems," *Proc. of the IEE*, Vol. 121, 1974, 424-429. DOI: 10.1049/piee.1974.0105 Cited on page(s) 8, 49

[72] Gongli, Z., Moraga, C., "Polynomial Fourier transforms," *Proc. 18th Int. Symp. on Multiple-Valued Logic*, May 24-26, 1988, 412-419. Cited on page(s) 110

[73] Guthrie, E., "The field of logic," *Journal of Philosophy and Scientific Methods*, Vol. 13, 1916, 152-158, and 336. DOI: 10.2307/2012970 Cited on page(s) 9

[74] Haar, A., "Zur theorie der orthogonalen Funktionsysteme," *Math. Annal.*, 69, 1910, 331- 371. DOI: 10.1007/BF01456326 Cited on page(s) 49

[75] Hansen, J.P., Sekine, M., "Synthesis by spectral translation using Boolean decision diagrams," *Proc. Design. Automation Conf.*, June 1996, 248-253. DOI: 10.1145/240518.240564 Cited on page(s) 93

[76] Hansen, J.P., Sekine, M., "Decision diagrams based techniques for the Haar wavelet transform," *Proc. IEEE Int. Conf. on Information, Communications and Signal Processing* (1st ICICS), Singapore, Vol. 1, September 1997, 59-63. DOI: 10.1109/ICICS.1997.647057 Cited on page(s) 93

[77] Harking, B., Moraga, C., "Efficient derivation of Reed-Muller expansions in multiple-valued logic systems," *Proc. 22nd Int. Symp. on Multiple-Valued Logic*, Sendai, Japan, May 27-29, 1992, 436-441. DOI: 10.1109/ISMVL.1992.186828 Cited on page(s) 8, 18, 25, 35

[78] Heidtmann, K.D., "Arithmetic spectrum applied to fault detection for combinational networks," *IEEE Trans. on Computers*, Vol. 40, No. 3, 1991, 320-324. DOI: 10.1109/12.76409 Cited on page(s) 17, 41

[79] Holowinski, G., "Parallel version of generalized Zakrevskij's algorithm for minimization of weakly specified multi-valued functions," *Proc. Int. Conf. on Pattern Recognition and Information Processing*, Minsk, Belarus, Vol. 1, 1997, 332-339. Cited on page(s) 48

[80] Holowinski, G., Yanushkevich, S., "Fast heuristic minimization of MVL functions in generalized Reed-Muller domain," *Proc. Int. Conf. on Applications of Computer Systems*, Szczecin, Poland, 1996, 57-64. Cited on page(s) 48

[81] Hong, Q., Fei, B., Wu, H., Perkowski, M.A., Zhuang, N., "Fast synthesis for ternary Reed-Muller expansion," *Proc. of The Twenty-Third International Symposium on Multiple-Valued Logic*, 1993, 14-16. DOI: 10.1109/ISMVL.1993.289588 Cited on page(s) 49

[82] Hurst, S.L., *Logical Processing of Digital Signals*, Crane Russak and Edward Arnold, London and Basel, 1978. Cited on page(s) 11, 93

[83] Hurst, S.L., "An engineering consideration of spectral transforms for ternary logic synthesis," *The Computer Journal*, Vol. 22, No. 2, 1979, 173-183. DOI: 10.1093/comjnl/22.2.173 Cited on page(s) 74

[84] Hurst, S.L., "The Haar transform in digital network synthesis," *Proc. 11th Int. Symp. on Multiple-valued Logic*, Oklahoma City, Oklahoma, USA, May 1981, 10-18. Cited on page(s) 93

[85] Hurst, S.L., Miller, D.M., Muzio, J.C., *Spectral Techniques in Digital Logic*, Academic Press, 1985. Cited on page(s) 8, 17

[86] Jabir, A.M.; Pradhan, D.K.; Mathew, J., $GfXpress$ - A Technique for synthesis and optimization of $GF(2^m)$ polynomials," *IEEE Trans. Computer-Aided Design of Integrated Circuits and Systems*, Vol. 27, No. 4, 2008, 698-711. DOI: 10.1109/TCAD.2008.917586 Cited on page(s) 27

[87] Janković, D., Stanković, R.S., Drechsler, R., "Tabular techniques for MV logic," in Soldek, J., Pejas, J., (eds.), *Advanced Computer Systems*, Kluwer Academic Publishers, 2002, 433-448. Cited on page(s) 27

[88] Janković, D., Stanković, R.S., Drechsler, R., "Efficient calculation of fixed-polarity polynomial expressions for multiple-valued logic functions," *Proc. 32nd Int. Symp. on Multiple-Valued Logic*, Boston, Massachusetts, USA, May 15-18, 2002, 76-82. DOI: 10.1109/ISMVL.2002.1011073 Cited on page(s) 27

[89] Janković, D., Stanković, R.S., Drechsler, R., "Decision diagrams optimization using copy properties," *EUROMICRO 2002*, Dortmund, Germany, September 2002. DOI: 10.1109/DSD.2002.1115374 Cited on page(s) 81

[90] Janković, D., Stanković, R.S., Drechsler, R., "Reduction of sizes of multi-valued decision diagrams by copy properties," *Proc. 34th Int. Symp. on Multiple-Valued Logic*, Toronto, Canada, May 19-22, 2004, 229-234. DOI: 10.1109/ISMVL.2004.1319945 Cited on page(s) 81

[91] Janković, D., Stanković, R.S., Moraga, C., "Optimization of Kronecker expressions using the extended dual polarity property," *ICIEST 2002*, October 4-6, 2002, 749-752. Cited on page(s) 27

[92] Janković, D., Stanković, R.S., Moraga, C., "Optimization of $GF(4)$ expressions using the extended dual polarity property," *Proc. 33rd Int. Symp. on Multiple-valued Logic*, May 16-19, 2003, 50-55. DOI: 10.1109/ISMVL.2003.1201384 Cited on page(s) 27

[93] Janković, D., Stanković, R.S., Moraga, C., "Arithmetic expressions optimization using dual polarity property," *Serbian Journal of Electrical Engineering*, Vol. 1, No. 1, November 2003, 71-80. DOI: 10.2298/SJEE0301071J Cited on page(s) 27

[94] Janković, D., Stanković, R.S., Moraga, C., "Exploiting homogeneous dual polarity routes in implementation of algorithms for optimization of Galois field expressions for ternary functions," *Proc. 37th Int. Symp. on Multiple-Valued Logic*, May 2007, 28-28 (CD-publication). DOI: 10.1109/ISMVL.2007.22 Cited on page(s) 27

[95] Janković, D., Stanković, R.S., Moraga, C., "Optimization of polynomial expressions by using the extended dual polarity," *IEEE Trans. Computers*, Vol. 58, No. 12, 2009, 1710-1725. DOI: 10.1109/TC.2009.113 Cited on page(s) 27

[96] Jaroszewicz, S., Shmerko, V., Yanushkevich, S., "Exact irredundant searching for a minimal Reed-Muller expansion for an incompletely specified MVL function," *Proc. Int. Conf. on Applications of Computer Systems*, Szczecin, Poland, 1996, 65-74. Cited on page(s) 48

[97] Kalganova, T., Kochergov, E., Zaitseva, E., Yanushkevich, S., "A genetic approach to optimise polynomial forms of incompletely specified MVL functions," *Proc. Workshop on Evolutionary Computing*, Brighton, UK, 1996, 89-102. Cited on page(s) 48

[98] Karpovsky, M.G., *Finite Orthogonal Series in the Design of Digital Devices*, John Wiley, 1976. Cited on page(s) 8, 16, 18, 109, 110, 111

[99] Karpovsky, M.G., Stanković, R.S., Astola, J.T., *Spectral Logic and Its Applications in the Design of Digital Devices*, Wiley, 2008. DOI: 10.1002/9780470289228 Cited on page(s) 11, 72, 93, 125

[100] Karpovsky, M.G., Stanković, R.S., Moraga, C., "Spectral techniques in binary and multiple-valued switching theory, a review of results in the previous decade," *Multiple-Valued Logic and Soft Computing*, Vol. 10, No. 3, 2004, 261-286. DOI: 10.1109/ISMVL.2001.924553 Cited on page(s) 74

[101] Kebschull, U., Schubert, E., Rosenstiel, W., "Multilevel logic synthesis based on functional decision diagrams," *Proc. 3rd European Conf. on Design Automation*, 1992, 43-47. DOI: 10.1109/EDAC.1992.205890 Cited on page(s) 78

[102] Kodandapani, K.L., Setur, R.V., "Multi-valued algebraic generalization of Reed-Muller canonical forms," *Proc. Int. Symp. on Multiple-Valued Logic*, 1974, 505-544. Cited on page(s) 8, 49

[103] Lai, Y.F., Pedram, M., Vrudhula, S.B.K., "EVBDD-based algorithms for integer linear programming, spectral transformation, and functional decomposition," *IEEE Trans. Computer-Aided Design of Integrated Circuits and Systems*, Vol.13, No. 8, 1994, 959-975. DOI: 10.1109/43.298033 Cited on page(s) 79, 99, 100, 102, 103, 106

[104] Lee, C.Y., Chen, W.H., "Several-valued combinational switching circuit," *Trans. AIEE*, Vol. 75, 1956, 278-283. Cited on page(s) 11

[105] Lee, S.C., *Modern Switching Theory and Digital Design*, Prentice-Hall, 1978. Cited on page(s) 8

[106] Lee, S. C., T. C. Lee, "On multi-valued symmetric functions," IEEE Trans. Computers, C-21, 312-316, March, 1972. DOI: 10.1109/TC.1972.5008957 Cited on page(s) 12

[107] Luis, M., Moraga, C., "On functions with flat Chrestenson spectra," *Proc. 19th Int. Symp. on Multiple-Valued Logic*, May 29-31, 1989, 406-413. DOI: 10.1109/ISMVL.1989.37814 Cited on page(s) 74

[108] Lukasiewicz, J., "O logice trójwartos'ciowej," *Ruch filozoficzny*, 5, 1920, 170–171, (in Polish). English translation, "On three-valued logic," in L. Borkowski (ed.), *Selected works by Jan Lukasiewicz*, North–Holland, Amsterdam, 1970, 87-88. Cited on page(s) 9

[109] Lukasiewicz, J., "Philosophical remarks on many-valued systems of propositional logic," in S. McCall (ed.), *Polish Logic, 1920 - 1939*, Oxford, 1967, 40-65. Cited on page(s) 9

[110] MacWilliams, S., *The Theory of Error-correcting Codes*, North-Holland, Amsterdam, 1977. Cited on page(s) 54, 56

[111] Malyugin, V.D., *Paralleled Calculation by Means of Arithmetic Polynomials*, Physical and Mathematical Publishing Company, Russian Academy of Science, Moscow, 1997. Cited on page(s) 8, 41, 42

[112] Malyugin, V.D., Sokolov, V.V., "Intensive logic computation," *Automatika and Telemekhanika*, 1993, No.4. Cited on page(s) 17, 41, 42

[113] Martin, N.M., "The Sheffer functions of 3-valued logic," *J. Symbolic Logic*, Vol. 19, No. 1, 1954, 45–51. DOI: 10.2307/2267650 Cited on page(s) 11

[114] Menger, K.S., "A transform for logic networks," *IEEE Trans. Computers*, Vol.C-18, No.3, 1969, 241-251. DOI: 10.1109/T-C.1969.222637 Cited on page(s) 25, 35, 36, 74

[115] Miller, D.M., "A canonical representation for many-valued symmetric functions," Proc. 6th Manitoba Conf. Numerical Mathematics and Computing, Oct. 1976, 303-313. Cited on page(s) 12

[116] Miller, D.M., "Spectral transformation of multiple-valued decision diagrams," *Proc. 24th Int. Symp. on Multiple-Valued Logic*, Boston, Massachusetts, USA, May 25-27, 1994, 89-96. DOI: 10.1109/ISMVL.1994.302209 Cited on page(s) 100

[117] Miller, D.M., Drechsler, R., "On the construction of multiple-valued decision diagrams," *Proc. 32nd Int. Symp. on Multiple-Valued Logic*, Boston, Massachusetts, USA, May 15-18, 2002, 245-253. DOI: 10.1109/ISMVL.2002.1011095 Cited on page(s) 106

[118] Miller, D.M., Muranaka, N., "Multiple-valued decision diagrams with symmetric variable nodes," *Proc. 26th Int. Symp. on Multiple-Valued Logic*, Santiago de Compostela, Spain, May 29-31, 1996, 242-247. DOI: 10.1109/ISMVL.1996.508375 Cited on page(s) 81, 100, 101

[119] Miller, D.M., Thorton, M.A., *Multiple Valued Logic - Concepts and Representations*, Morgan & Calypool, 2008. DOI: 10.2200/S00065ED1V01Y200709DCS012 Cited on page(s) 11

[120] Minato, S., "Graph-based representations of discrete functions," in Sasao, T., Fujita, M., (Ed.), *Representations of Discrete Functions*, Kluwer Academic Publishers, 1996, 1-28. DOI: 10.1007/978-1-4613-1385-4 Cited on page(s) 79

[121] Minasyan, S., Astola, J., Egiazarian, K., Stanković, R.S., "Hybrid Reed-Muller-Haar transform and its applications in reduction the spectral representations of logic functions," *Proc. 38th Int. Symp. on Multiple-Valued Logic*, Dallas, Texas, USA, May 22-24, 2008, 232-237. DOI: 10.1109/ISMVL.2008.8 Cited on page(s) 49

[122] Minasyan, S., Stanković, R.S., Astola, J., "Ternary Haar-like transform and its application in spectral representation of ternary-valued functions," *Lecture Notes in Computer Science*, 2009, Vol. 5717, PGS, 518-525, ISBN 978-3-642-04771-8, ISSN 0302-9743, ISSN2 1611-3349. DOI: 10.1007/978-3-642-04772-5_67 Cited on page(s) 49

[123] Minasyan, S., Stanković, R.S., Egiazarian, K., Astola, J., "Hybrid Reed-Muller Haar representations of logic functions," *Multiple-Valued Logic and Soft Computing*, Vol. 15, No. 4, 2009, 341-359. DOI: 10.1109/ISMVL.2008.8 Cited on page(s) 49

[124] Moisil, G., "Recherches sur l'algèbra de la logique," *Ann. Sci. Univ. Jassy*, 22, 1935, 1-117. Cited on page(s) 9

[125] Moisil, G, "Recherches sur les logiques non-Chrysippiennes," *Annalles Scientifiques de l' Udversité de Jassy*, premiere section, 26, 1940, 431-466. Cited on page(s) 9

[126] Moisil, G.C., "Notes sur les logiques non-chrysippiennes," *Ann. Sci. Univ. Jassy*, 27, 86-98, 1941. Cited on page(s) 9

[127] Moisil, G., *Essays sur les Logiques Nonchrysippiénnes*, Académie de la R'epublique Socialiste de Roumanie, Bucharest, 1972. Cited on page(s) 9

[128] Moraga, C., "Ternary spectral logic," *Proc. Int. Symp. on Multiple-Valued Logic*, 1977, 7-12. Cited on page(s) 74

[129] Moraga, C., "Complex spectral logic," *Proc. 8th Int. Symp. on Multiple-valued Logic*, Rosemond, Illinois, U.S.A., 1978, 149-156. Cited on page(s) 18, 74

[130] Moraga, C., "Introducing disjoint spectral translation in spectral multiple-valued logic design," *Electronics Letters*, Vol. 14, No. 8, 1978, 241-243. DOI: 10.1049/el:19780164 Cited on page(s) 74

[131] Moraga, C., "Spectral characterisation of ternary threshold functions," *Electronics Letters*, Vol. 15, No. 12, 1979, 712-713. DOI: 10.1049/el:19790573 Cited on page(s) 74

[132] Moraga, C., "Characterisation of ternary threshold functions using a partial spectrum," *Electronics Letters*, Vol. 15, No. 24, 1979, 803-805. DOI: 10.1049/el:19790573 Cited on page(s) 74

[133] Moraga, C., "Introduction to linear p-adic invariant systems," In: *Cybernetics and Systems Research*, Vol. 2, 121-124, (Ed.), R. Trappl, Vienna: Electronic Science Publ., 1984. Cited on page(s) 74

[134] Moraga, C., "On some applications of the Chrestenson functions in logic design and data processing," Mathematics and Computers in Simulation, Vol.27, No.5-6, 1985, 431-439. DOI: 10.1016/0378-4754(85)90062-X Cited on page(s) 18, 74

[135] Moraga, C., "Design of a multiple-valued systolic system for the computation of the Chrestenson spectrum," *IEEE Trans. Computers*, Vol. C-35, No. 2, 1986, 183-188. DOI: 10.1109/TC.1986.1676739 Cited on page(s) 74

[136] Moraga, C., "A decade of spectral techniques," *Proc. 21st Int. Symp. on Multiple-Valued Logic*, May 26-29, 1991, 182-288. DOI: 10.1109/ISMVL.1991.130726 Cited on page(s) 74

[137] Moraga, C., "Improving the characterization of p-valued threshold functions," *Proc. 32nd Int. Symp. on Multiple-Valued Logic*, May 15-18, 2002, 28-34. DOI: 10.1109/ISMVL.2002.1011066 Cited on page(s) 74

[138] Moraga, C., Oenning, R., "The 2D Zhang-Watari orthogonal transform," No. 567 of Forschungsberichte des Fachbereichs Informatik der Universität Dortmund, Universität (Dortmund), Germany, Fachbereich Informatik, Publisher Dekanat Informatik, Univ., 1995. Cited on page(s) 110

[139] Moraga, C., Oenning, R., Karpovsky, M.G., "The Zhang-Watari transform - A discrete, real-valued, generalized Haar transform," *Multi Valued Logic*, Vol. 2, 1997, 245-262. Cited on page(s) 110

[140] Moraga, C., Stanković, R.S., Astola, J.T., "Properties of matrix-valued spectral coefficients obtained with the Fourier Transform on a non-Abelian group," *Proc. 36th Int. Symp. on Multiple-Valued Logic*, May 17-20, 2006, Singapore, 35/1-35/6. DOI: 10.1109/ISMVL.2006.34 Cited on page(s) 78

[141] Moraga, C., Stanković, M., Stojković, S., "Spectral analysis of special properties of ternary functions," *Proc. 37th Int. Symp. on Multiple-Valued Logic*, May 2007, 4-4 (CD-publication). DOI: 10.1109/ISMVL.2007.52 Cited on page(s) 12

[142] Muller, D.E., "Application of Boolean algebra to switching circuits design and to error detection," *IRE Trans. Electron. Comp.*, Vol. EC-3, 1954, 6-12. Cited on page(s) 15

[143] Mukhopadhyay, A., "Symmetric ternary switching functions," IEEE Trans. Computers., EC-15, 731-735, 1966. DOI: 10.1109/PGEC.1966.264561 Cited on page(s) 12

[144] Muzio, J.C., "Stuck fault sensitivity of Reed-Muller and Arithmetic coefficients," C. Moraga, Ed., *Theory and Applications of Spectral Techniques*, Dortmund, 1989, 36-45. Cited on page(s) 17, 41

[145] Muzio, J.C., Wesselkamper, T.C., *Multiple-Valued Switching Theory*, Adam Hilger, Bristol, 1986. Cited on page(s) 8, 9, 11, 17, 25, 35, 83, 86

[146] Nagayama, S., Sasao, T., Butler, J.T., "Numeric function generators using piecewise arithmetic expressions," *Proc. 41th Int. Symp. on Multiple-Valued Logic*, Tuusula, Finlad, May 23-25, 2011, 16-21. DOI: 10.1109/ISMVL.2011.32 Cited on page(s) 54

[147] Nagayama, S., Sasao, T., Butler, J.T., "A systematic design method for two-variable numeric function generators using multiple-valued decision diagrams," *IEICE Trans. Information and Systems*, Vol. E93-D, No. 8, 2010, 2059-2067. DOI: 10.1587/transinf.E93.D.2059 Cited on page(s) 54

[148] Oenning, R., Moraga, C., "Properties of the Zhang-Watari transform," *Proc. 25th Int. Symp. on Multiple-Valued Logic*, May 23-25, 1995, 44-49. DOI: 10.1109/ISMVL.1995.513508 Cited on page(s) 110

[149] Peirce, C.S., *Collected Papers*, Cambridge, Massachusetts, USA, 1933, Vols. 3-4. Cited on page(s) 9

[150] Post, E.L., "Introduction to a general theory of elementary propositions," *Amer. J. Math.*, Vol. 43, 1921, 163-185. DOI: 10.2307/2370324 Cited on page(s) 9, 10

[151] Post, E.L., *The two-valued iterative systems of mathematical logic*, Princeton Univ. Press, Princeton, N.J., 1941. Cited on page(s) 9, 10

[152] Pradhan, D.K., "A multi-valued algebra based on finite fields," *Proc. 1974 Int. Symp. on Multiple-Valued Logic*, Morgantown, WV, USA, May 1974, 95-112. Cited on page(s) 8, 11

[153] Pradhan, D.K., "A Theory of Galois switching functions," *IEEE Trans. Computers*, Vol. C-27, No. 3, 1978, 239-248. DOI: 10.1109/TC.1978.1675077 Cited on page(s) 11

[154] Rahardja, S., Falkowski, B.J., "Family of fast mixed arithmetic Logic transforms for multiple-valued input binary functions," *Proc. 26th Int. Symp. on Multiple-Valued Logic*, 1996, 24-29. DOI: 10.1109/ISMVL.1996.508331 Cited on page(s) 54

[155] Rahardja, S., Falkowski, B.J., "Fast Linearly independent arithmetic expansions," *IEEE Trans. Computers*, Vol. 48, No. 9, 1999, 991-999. DOI: 10.1109/12.795227 Cited on page(s) 54

[156] Rahardja, S., Falkowski, B.J., "A new algorithm to compute quaternary Reed-Muller expansions," *Proc. 30th IEEE International Symposium on Multiple-Valued Logic*, 2000, 153-158. DOI: 10.1109/ISMVL.2000.848614 Cited on page(s) 17, 49

[157] Rahardja, S., Falkowski, B.J., "Fast mixed linearly independent arithmetic logic transforms for multiple-valued functions," *Mult. Val. Logic and Soft Computing*, Vol. 10, No. 1, 2004, 73-87. DOI: 10.1109/12.795227 Cited on page(s) 17, 54

[158] Reed, I.S., " A class of multiple-error-correcting circuits and their decoding scheme," *IRE Trans. Inform. Theory*, PGIT-4, 1954, 38-49. DOI: 10.1109/TIT.1954.1057465 Cited on page(s) 15

[159] Rine, D.C., "Multiple-valued logic and computer science in 20th century," *IEEE Computer*, Vol. 7, 1975, 18-19. Cited on page(s) 11

[160] Rine, D.C., "An introduction to multiple-valued logic," in Rine, D.C., (ed.), *Computer Science and Multiple-Valued Logic Theory and Applications*, North-Holland Publishing Company, Amsterdam, The Netherlands, 1977. Cited on page(s) 11

[161] Rudin, W., *Fourier Analysis on Groups*, Interscience Publisher, New York, 1960. Cited on page(s) 71

[162] Salomaa, A., "On many-valued systems of logic," *Ajatus*, No. 22, 1959, 115-159. Cited on page(s) 9

[163] Sarabi, A., Perkowski, M.A., "Fast exact and quasi-minimal minimization of highly testable fixed polarity AND/XOR canonical networks," *Proc. Design Automation Conference*, June 1992, 30-35. DOI: 10.1109/DAC.1992.227867 Cited on page(s) 12

[164] Sasao, T., "Input-variable assignment and output phase optimization of programmable logic arrays," *IEEE Trans. Computers*, Vol. C-33, 1984, 879-894. DOI: 10.1109/TC.1984.1676349 Cited on page(s) 5, 7

[165] Sasao, T., "Optimization of multiple-valued AND-EXOR expressions using multiple-place decision diagrams," *Proc. 22nd IEEE Int. Symp. on Multiple-Valued Logic*, 1992, 451-458. DOI: 10.1109/ISMVL.1992.186830 Cited on page(s) 49

[166] Sasao, T., "AND-EXOR expressions and their optimization" in T. Sasao, (ed.), *Logic Synthesis and Optimization*, Kluwer Academic Publishers, 1993. DOI: 10.1007/978-1-4615-3154-8 Cited on page(s) 16, 66

[167] Sasao, T., *Switching Theory for Logic Synthesis*, Kluwer Academic Publishers, 1999. DOI: 10.1007/978-1-4615-5139-3 Cited on page(s) 5, 7, 16

[168] Sasao, T., "Arithmetic ternary decision diagrams and their applications," *Fourth International Workshop on Applications of the Reed-Muller Expansion in Circuit Design, (Reed-Muller 99)*, Victoria, Canada, August 20-21, 1999. Cited on page(s) 42

[169] Sasao, T., Butler, J.T., "A design method for look-up table type FPGA by pseudo-Kronecker expansions," *Proc. 24th Int. Symp. on Multiple-valued Logic*, Boston, Massachusetts, 25.-27.5. 1994, 97-104. DOI: 10.1109/ISMVL.1994.302215 Cited on page(s) 5, 7, 49, 100

[170] Sasao, T., Fujita, M., (Eds.), *Representations of Discrete Functions*, Kluwer Academic Publishers, 1996. DOI: 10.1007/978-1-4613-1385-4 Cited on page(s) 78, 79, 80, 87, 99, 105, 124

[171] Sasao, T., Butler, J.T., "Comparison of the worst and best sum-of-products expressions for multiple-valued functions," *Proc. 27th Int. Symp. on Multiple-Valued Logic*, Nova Scotia, Canada, May 28-30, 1997, 55-60. DOI: 10.1109/ISMVL.1997.601374 Cited on page(s) 12

[172] Schafer, I., Perkowski, M.A., "Multiple-valued generalized Reed-Muller forms," *Proc. of the Twenty-First International Symposium on Multiple-Valued Logic*, 1991, 40-48. DOI: 10.1109/ISMVL.1991.130703 Cited on page(s) 49

[173] Schafer, I., Perkowski, M.A., "Multiple valued input generalised Reed-Muller forms," *IEE Proceedings Computers and Digital Techniques*, Part E, Vol. 139, No. 6, 1992, 519-527. DOI: 10.1049/ip-e.1992.0074 Cited on page(s) 49

[174] Shmerko, V., Holowinski, G., Song, N., Dill, K., Oanguly, K., Salranek, R., Perkowski, M., "High-quality minimization of multi-valued input binary-output Exclusive-Or Sum of Product expressions for strongly unspecified multi-output functions," *Proc. Int. Conf. on Applications of Computer Systems*, Poland, 1997, 248-255. Cited on page(s) 49

[175] Song, N., Perkowski, M.A., "EXORCISM-MV-2: Minimisation of Exclusive Sum of Products expressions for multiple-valued input incompletely specified functions," *Proc. 23rd Int. Symp. on Multiple-Valued Logic*, 1993, 132-137. DOI: 10.1109/ISMVL.1993.289569 Cited on page(s) 49

[176] Song, N., Perkowski, M., "Minimization of Exclusive Sum of Products expressions for multi-output multiple-valued input, incompletely specified functions," *IEEE Trans. on CAD*, Vol. 15, No. 4, 1996, 385-395. DOI: 10.1109/ISMVL.1993.289569 Cited on page(s) 49

[177] Srinivasan, A., Kam, T., Malik, Sh., Brayant, R.K., "Algorithms for discrete function manipulation," in: *Proc. Inf. Conf. on CAD*, 1990, 92-95. DOI: 10.1109/ICCAD.1990.129849 Cited on page(s) 81, 106, 107

[178] Stanković, M., Janković, D., Stanković, R.S., "Efficient algorithm for Haar spectrum calculation," *Scientific Review*, No. 21-22, 1996, 171-182. Cited on page(s) 125

[179] Stanković, M., Janković, D., Stanković, R.S., "Efficient algorithm for Haar spectrum calculation," *Proc. IEEE Int. Conf. on Information, Communications and Signal Processing* (1st ICICS), Singapore, Vol. 4, September 1997, 6-10. Cited on page(s) 125

[180] Stanković, M., Janković, D., Stanković, R.S., Falkowski, B.J., "Calculation of the paired Haar transform through shared binary decision diagrams," *Computers and Electrical Engineering*, Vol. 29, No. 1, 2003, 13-24. DOI: 10.1016/S0045-7906(01)00022-2 Cited on page(s) 125

[181] Stanković, M., Stojković, S., "Calculation of symmetric transform of Boolean functions represented by decision diagrams," *Proc. XLII Yugoslav Conference for ETRAN*, Vrnjačka Banja, June 3-5, 1998, 63-66, (in Serbian). Cited on page(s) 125

[182] Stanković, M., Stojković, S., Moraga, C., "Linearization of ternary decision diagrams by using the polynomial Chrestenson spectrum," *Proc. 37th Int. Symp. on Multiple-Valued Logic*, May 2007, 41-41 (CD-publication). DOI: 10.1109/ISMVL.2007.31 Cited on page(s) 74

[183] Stanković, R.S., "Some remarks on the canonical forms for pseudo-Boolean functions," *Publ. Inst. Math. Beograd*, (N.S.), 37, 51, 1985, 3-6. Cited on page(s) 16

[184] Stanković, R.S., "Some remarks on Fourier transforms and differential operators for digital functions," *Proc. 22nd Int. Symp. on Multiple-Valued Logic*, Sendai, Japan, 1992, 365-370. DOI: 10.1109/ISMVL.1992.186818 Cited on page(s) 8, 11, 18, 28, 29

[185] Stanković, R.S., "Some remarks about spectral transform interpretation of MTBDDs and EVBDDs", *Proc. of the Asian and South Pacific Design Automation Conference*, (ASP-DAC'95), 29.8-1.9.1995, Makuhari Messe, Chiba, Japan, 1995, 385-390. DOI: 10.1109/ASPDAC.1995.486348 Cited on page(s) 100, 103

[186] Stanković, R.S., "Functional decision diagrams for multiple-valued functions," *Proc. 25-th Int. Symp. on Multiple-Valued Logic*, 23-25.5.1995, Bloomington, Indiana, U. S. A., 284-289. DOI: 10.1109/ISMVL.1995.513544 Cited on page(s) 84, 86, 100

[187] Stanković, R.S., "Edge-valued decision diagrams based on partial Reed-Muller transforms," *Proc. Reed-Muller Colloquium UK'95*, Bristol, England, UK, December 19, 1995, 9/1-9/13. Cited on page(s) 100, 102, 105

[188] Stanković, R., "Fourier decision diagrams on finite non-Abelian groups with preprocessing," *ISMVL-27*, May 1997, 281-286. DOI: 10.1109/ISMVL.1997.601415 Cited on page(s) 7, 78

[189] Stanković, R.S., "Functional decision diagrams for multi-valued functions," *Multi. Val. Logic*, Vol. 3, 1998, 195-215. DOI: 10.1109/ISMVL.1995.513544 Cited on page(s) 84

[190] Stanković, R.S., "Non-Abelian groups in optimization of decision diagrams representations of discrete functions," *Formal Methods in System Design*, Vol. 18, 2001, 209-231. DOI: 10.1023/A:1011265018200 Cited on page(s) 7

[191] Stanković, R.S., "Unified view of decision diagrams for representation of discrete functions," *Multi. Val. Logic*, Vol. 8, No. 2, 2002, 237-283. DOI: 10.1080/10236620215293 Cited on page(s) 7, 78, 81, 89

[192] Stanković, R.S., "Arithmetic transform ternary decision diagrams for exact minimization of fixed polarity arithmetic expressions," *Multiple-Valued Logic and Soft Computing*, Vol. 10, No. 3, 2004, 287-307. Cited on page(s) 42

[193] Stankovic, R.S., Astola, H., Astola, J.T., "Determining minimized Galois field expressions for ternary functions," *Proc. 41st Int. Symp. on Multiple-Valued Logic*, Tuusula, Finland, May 23-25, 2011, 117-124. DOI: 10.1109/ISMVL.2011.26 Cited on page(s) 49

[194] Stanković, R.S., Astola, J.T., "Design of decision diagrams with increased functionality of nodes through group theory," *IEICE Trans. Fundamentals*, Vol. E86-A, No. 3, 2003, 693-703. Cited on page(s) 78

[195] Stanković, R.S., Astola, J.T., *Spectral Interpretation of Decision Diagrams*, Springer, 2003. Cited on page(s) 16, 77, 78, 99, 100, 101, 117

[196] Stanković, R.S., Astola, J.T., "Edge-valued decision diagrams for multiple-valued functions," *Proc. 34th Int. Symp. on Multiple-Valued Logic*, Toronto, Canada, May 19-22, 2004, 229-234. DOI: 10.1109/ISMVL.2004.1319946 Cited on page(s) 105

[197] Stanković, R.S., Astola, J., "Remarks on the complexity of arithmetic representations of elementary functions for circuit design," *Workshop on Applications of the Reed-Muller Expansion*

in Circuit Design and Representations and Methodology of Future Computing Technology, Oslo, Norway, May 2007, 5–11. Cited on page(s) 54

[198] Stanković, R.S., Astola, J.T., Egiazarian, K., "Remarks on symmetric binary and multiple-valued logic functions," *Proc. 6th Int. Workshop on Boolean Problems*, Freiberg, Germany, September 23-24, 2004, 83-87. Cited on page(s) 12

[199] Stanković, R.S., Astola, J.T., Egiazarian, K., "Recent developments in Haar wavelet transform for application to switching and multivalued logic functions representations," in *Advances in Signal Transforms, Theory and Applications*, J. Astola, L. Yaroslavsky, (eds.), EURASIP Book Series on Signal Processing and Communications, Vol. 7, Hindawi Publishing Corporation, 2007, 57-92. Cited on page(s) 51, 93, 94

[200] Stanković, R.S., Drechsler, R., "Circuit design from Kronecker Galois field decision diagrams for multiple-valued functions," *ISMVL-27*, May 1997, 275-280. DOI: 10.1109/ISMVL.1997.601413 Cited on page(s) 87

[201] Stanković, R.S., Janković, D., Moraga, C., "Reed-Muller-Fourier versus Galois field representations of four-valued logic functions," *Proc. 28th Int. Symp. on Multiple-Valued Logic*, May 27-29, 1998, 186-191. DOI: 10.1109/ISMVL.1998.679340 Cited on page(s) 35

[202] Stanković, R.S., Moraga, C., "Fast algorithms for detecting some properties of multiple-valued functions, *Proc. 14th Int. Symp. on Multiple-valued Logic*, Winnipeg, Canada, May 29-31, 1984. Cited on page(s) 12

[203] Stanković, R.S., Moraga, C., "Methods for the detection of some properties of multiple-valued functions," *IEE Proc.*, Part E, Vol.139, No.5, 1992, 421-429. Cited on page(s) 12

[204] Stanković, R.S., Janković, D., Moraga, C., "Reed-Muller-Fourier versus Galois field representations of four-valued logic functions"*Proc. 28th IEEE International Symposium on Multiple-Valued Logic*, 1998, 186-191. DOI: 10.1109/ISMVL.1998.679340 Cited on page(s) 35, 46

[205] Stanković, R.S., Moraga, C., "Reed-Muller-Fourier representations of multiple-valued functions over Galois fields of prime cardinality," Kebschull, U., Schubert, E., Rosentiel, W., (eds.), *Proc. IFIP WG10.5 Workshop on Applications of the Reed-Muller Expansion in Circuit Design*, Hamburg, Germany, September 16-17, 1993, 115-124. Cited on page(s) 8, 11, 18, 28, 88

[206] Stanković, R.S., Stanković, M., Moraga, C., Sasao, T., "Calculation of Vilenkin-Chrestenson transform coefficients of multiple-valued functions through multiple-place decision diagrams," *Proc. 5th Int. Workshop on Spectral Techniques*, 15.-17.3.1994, Beijing, China, 107-116. Cited on page(s) 17, 18

[207] Stanković, R.S., Moraga, C., "An algebraic transform for prime-valued functions," *Proc. 5th Int. Workshop on Spectral Techniques*, 15.-17.3.1994, Beijing, China, 205-209. Cited on page(s) 46

[208] Stanković, R.S., Moraga, C., "Edge-valued functional decision diagrams," *Proc. Int. Conf. on Computer Aided Design of Discrete Devices*, 15.-17.11.1995, Minsk, Belarus, Vol. 2, 69-73. Cited on page(s) 103

[209] Stanković, R.S., Moraga, C., "Reed-Muller-Fourier representations of multiple-valued functions," in Stanković, R.S., Stojić, M.R., Stanković, M.S., (eds.) *Recent Developments in Abstract Harmonic Analysis with Applications in Signal Processing*, Nauka, Belgrade, Elektronski fakultet, Niš, 1996, 205-216. Cited on page(s) 7, 18, 28

[210] Stankovć, R.S., Moraga, C., "Edge-valued decision diagrams for multiple-valued functions," *Proc. 3rd Int. Conf. Application of Computer Systems*, Szczecin, Poland, November 21-22, 1996. DOI: 10.1109/ISMVL.2004.1319946 Cited on page(s) 100, 102, 104, 105

[211] Stanković, R.S., Moraga, C., Astola, J.T., "From Fourier expansions to arithmetic-Haar expressions on quaternion groups," *Applicable Algebra in Engineering, Communication and Computing*, Vol. AAECC 12, 2001, 227-253. DOI: 10.1007/s002000100068 Cited on page(s) 49

[212] Stanković, R.S., Moraga, C., Astola, J.T., "Reed-Muller expressions in the previous decade," *Multiple-Valued Logic and Soft Computing*, Vol. 10, No. 1, 2004, 5-28. Cited on page(s) 78, 89

[213] Stanković, R.S., Moraga, C., Astola, J.T., "Derivatives for multiple-valued functions induced by Galois field and Reed-Muller-Fourier expressions," *Proc. 34th Int. Symp. on Multiple-Valued Logic*, Toronto, Canada, May 19-22, 2004, 184-189. DOI: 10.1109/ISMVL.2004.1319939 Cited on page(s) 16

[214] Stanković, R.S., Moraga, C., Astola, J.T., "Remarks on the structure of matrix-valued spectral transforms on finite non-Abelian groups," *Proc. 34th Int. Symp. on Multiple-Valued Logic*, Calgary, Canada, May 18-21, 2005, 188-193. DOI: 10.1109/ISMVL.2005.42 Cited on page(s) 7

[215] Stanković, R.S., Moraga, C., Astola, J.T., *Applications of Fourier Analysis on Finite Non-Abelian Groups in Signal Processing and System Design*, IEEE Press/Wiley, 2005. DOI: 10.1002/047174543X Cited on page(s) 7, 69

[216] Stanković, R.S., Sasao, T., "Decision Diagrams for Discrete Functions: Classification and Unified Interpretation," *ASP-DAC'98*, February 1998. DOI: 10.1109/ASPDAC.1998.669518 Cited on page(s) 94

[217] Stanković, R.S., Sasao, T., "Komamiya equation for multiple-valued adders," Invited talk, *Proc. 10th Int. Workshop on Post-Binary ULSI Systems*, Warszawa, Poland, May 21, 2001, 63-68. Cited on page(s) 12

[218] Stanković, R.S., Stanković, M., Astola, J.T., Egiazarian, K., "Haar spectral transform deci-sion diagrams with exact algorithm for minimization of the number of paths," *Proc. 4th Int. Workshop on Boolean Problems*, Freiberg, Germany, September 21-22, 2000. Cited on page(s) 93

[219] Stanković, R.S., Stanković, M., Janković, D., *Spectral Transforms in Switching Theory, Defini-tions and Calculations*, Nauka, Belgrade, 1998. Cited on page(s) 125

[220] Stanković, R.S., Stanković, M., Moraga, C., "Design of Haar wavelet transforms and Haar spectral transform decision diagrams for multiple-valued functions," *Proc. 31st Int. Symp. on Multiple-Valued Logic*, Warzsawa, Poland, May 22-24, 2001, 311-316. DOI: 10.1109/ISMVL.2001.924589 Cited on page(s) 49, 94, 108

[221] Stanković, R.S., Stanković, M., Moraga, C., Sasao, T., "Calculation of Reed-Muller-Fourier coefficients of multiple-valued functions through multiple-place decision diagrams," *Proc. Twenty-Fourth International Symposium on Multiple-Valued Logic*, Boston, Massachusetts, USA, May 22-25, 1994, 82-88. DOI: 10.1109/ISMVL.1994.302216 Cited on page(s) 124

[222] Stanković, R.S., Stanković, M., Moraga, C., Sasao, T., "Calculation of Vilenkin-Chrestenson transform coefficients of multiple-valued functions through multiple-place decision dia-grams," *Proc. 5th Int. Workshop on Spectral Techniques*, 15.-17.3.1994, Beijing, China, 107-116. Cited on page(s) 18, 124

[223] Steinbach, B., "Most complex Boolean functions detected by the specialized normal form," *Facta Universitatis, Ser., Elect. Engr.*, Vol. 20, No. 3, 2007, 259-279. DOI: 10.2298/FUEE0703259S Cited on page(s) 49

[224] Stojmenović, I., Miyakawa, M., Tošić, R., "On spectra of many-valued logic symmetric func-tions," *18th Int. Symp. on Multiple-Valued Logic*, 1988, 285-292. DOI: 10.1109/ISMVL.1988.5185 Cited on page(s) 12

[225] Tamari, D., "Some mutual applications of logic and mathematics," *Proc. 2nd Int. Colloq. Math. Logic*, August 1952, 89-90. Cited on page(s) 11

[226] Tan, E.C., Chia, C.Y., "Alternative algorithm for optimisation of Reed-Muller universal logic module networks," *IEE Proc. Computers and Digital Techniques*, Vol. 143, No. 6, 1996, 385-390. DOI: 10.1049/ip-cdt:19960770 Cited on page(s) 27

[227] Tan, E.C., Yang, H., "Optimization of Fixed-polarity Reed-Muller circuits using dual-polarity property," *Circuits, Systems, Signal Processing*, Vol. 19, No. 6, 2000, 535-548. DOI: 10.1007/BF01271287 Cited on page(s) 27

[228] Thayse, A., *Boolean Differential Calculus*, Springer-Verlag, 1980. Cited on page(s) 16

[229] Thayse, A., Davio, M., Deschamps, J.-P., "Optimization of multiple-valued decision diagrams," *Proc. 8th Int. Symp. on Multiple-Valued Logic*, 1978, 171-177. Cited on page(s) 81, 100

[230] Aiken, H., (ed.), *The Annals of the Computation Laboratory of Harvard University*, Volume XXVII, *Synthesis of Electronic Computing and Control Circuits*, Cambridge, Massachusetts, USA, 1951. Cited on page(s) 54

[231] Thornton, M.A., "Spectral transforms of mixed-radix MVL functions," *33rd Int. Symp. on Multiple-Valued Logic*, May 16-19, 2003, 329-333. DOI: 10.1109/ISMVL.2003.1201425 Cited on page(s) 27

[232] Thornton, M.A., "Mixed-radix MVL function spectral and decision diagram representation," *Automation and Remote Control*, Vol. 65, No. 6, 2004, 1007-1017. DOI: 10.1023/B:AURC.0000030910.23047.2d Cited on page(s) 27

[233] Thornton, M.A., Drechsler, R., "Computation of spectral information from logic netlists," *30th Int. Symp. on Multiple-Valued Logic*, May 23-25, 2000, 53-58. DOI: 10.1109/ISMVL.2000.848600 Cited on page(s) 118

[234] Thornton, M.A., Miller, D.M., "Computation of discrete function chrestenson spectrum using Cayley color graphs," *Multiple-Valued Logic and Soft Computing*, Vol. 10, No. 2, 2004, 189-202. DOI: 10.1109/ISMVL.2002.1011079 Cited on page(s) 118

[235] Thornton, M.A., Miller, D.M., Townsend, W.J., "Chrestenson spectrum computation using Cayley color graphs," *Proc. 32nd Int. Symp. on Multiple-Valued Logic*, Boston, Massachusetts, USA, May 15-18, 2002, 123-128. DOI: 10.1109/ISMVL.2002.1011079 Cited on page(s) 118

[236] Tošić, Ž., "Arithmetical representations of logic functions," in *Discrete Automatics and Connection Networks*, USSR Academy of Sciences, Nauka, Moscow, 1970, 131-136, (In Russian). DOI: 10.1201/9781420037586.ch5 Cited on page(s) 48

[237] Tošić, Ž., *Analytical Representation of m-Valued Logical Functions over the Ring of Integers Modulo m*, Ph.D. thesis, *Publ. Eect. Fac.*, Belgrade, 1972. Cited on page(s) 48

[238] Tošić, R., Stojmenović, I., Miyakawa, M., "On the maximum size of the terms in the realization of symmetric functions," *Proc. 21st Int. Symp. on Multiple-Valued Logic*, Victoria, BC, Canada, May 26-29, 1991, 110-117. Cited on page(s) 12

[239] Varma, D., Trachtenberg, E.A., "Efficient spectral techniques for logic synthesis," in: T. Sasao, Ed., *Logic Synthesis and Optimization*, Kluwer Academic Publishers, Boston, 1993, 215-232. DOI: 10.1007/978-1-4615-3154-8 Cited on page(s) 118

[240] Trachtenberg, E.A., "Applications of Fourier analysis on groups in engineering practices," in *Recent Developments in Abstract Harmonic Analysis with Applications in Signal Processing*, Nauka, Belgrade, 1996, 331-403. Cited on page(s) 118

[241] Varma, D., Trachtenberg, E.A., "Computation of Reed-Muller expansions of incompletely specified Boolean functions from reduced representations," *IEE Proceedings Computers and Digital Techniques*, Part E, Vol. 138, No. 2, 1991, 85-92. DOI: 10.1049/ip-e.1991.0011 Cited on page(s) 118

[242] Vasiliev, N.A., "Imaginary (non-Aristotelian) logic," *Atti del Congresso Internazionale di Filosofia*, (Napoli 5-9 Maggio 1924), 107-109. DOI: 10.1007/BF02229852 Cited on page(s) 9

[243] Vranesic, Z.G., Lee, E.S., Smith, K.C., "A many-valued algebra for switching systems," *IEEE Trans. Comput.*, Vol. C-19, 1970, 964-971. DOI: 10.1109/T-C.1970.222803 Cited on page(s) 11

[244] Vrudhula, S.B.K., Lai, Y.T., Pedram, M., "Efficient computation of the probability and Reed-Muller spectra of Boolean functions using edge-valued binary decision diagrams," in: Sasao, T., Fujita, M., Eds.,*Proc. IFIP WG 10.5 Workshop on Applications of the Reed-Muller Expansion in Circuit Design, Reed-Muller'95*, 27-29.8.1995, Makuhari, Chiba, Japan, 62-69. Cited on page(s) 100

[245] Watari, C., "A generalization of Haar functions," *Tohoky Math. Jr.*, 8, 1956, 286-290. DOI: 10.2748/tmj/1178244953 Cited on page(s) 110

[246] Webb, D.L., "Generation of any N-valued logic by one binary operator," *Proc. Nat. Acad. Sci.*, Vol. 21, 1935, 252-254. DOI: 10.1073/pnas.21.5.252 Cited on page(s) 11

[247] Wesselkamper, T.C., "Divided difference method for Galois switching functions," *IEEE Trans. Computers*, Vol.C-27, No.3, 1978, 232-238. DOI: 10.1109/TC.1978.1675076 Cited on page(s) 25, 35, 36

[248] Yanushkevich, S., "Matrix algorithms of synthesis of polynomial forms for MVL functions," *Proc. Int. Conf. on Parallel Processing and Applied Mathematics*, Poland, 1994, 113-122. Cited on page(s) 41

[249] Yanushkevich, S., "Spectral and differential methods to synthesize polynomial forms of MVL functions on systolic arrays," *Proc. 5th Int. Workshop on Spectral Techniques*, C. Moraga, Q. Zhang, (Eds.), Beijing, China, 1994, 78-93. Cited on page(s) 48

[250] Yanushkevich, S.N., "Matrix method for solving multivalued logic differential equations," *IEE Proc. Computers and Digital Techniques*, Vol. 144, No. 5, 1997, 267-272. DOI: 10.1049/ip-cdt:19971368 Cited on page(s) 8, 17, 48

[251] Yanushkevich, S.N., *Logic Differential Calculus in Multi-Valued Logic Design*, Techn. University of Szczecin Academic Publisher, Poland, 1998. Cited on page(s) 8, 16, 48

[252] Yanushkevich, S.N., Butler, J.T., Dueck, G.W., Shmerko, V.P., "Experiments on FPRM expressions for partially symmetric logic functions," *Proc. 30th IEEE Int. Symp. on Multiple-Valued Logic*, 2000, 141-146. DOI: 10.1109/ISMVL.2000.848612 Cited on page(s) 12

[253] Yanushkevich, S., Holowinski, G., "Fast heuristic minimization of MVL functions in generalized Reed-Muller domain," *Proc. Int. Conf. on Applications of Computer Systems*, Szczecin, Poland, 1996, 57-6.1. Cited on page(s) 49

[254] Yanushkevich, S.N., Miller, D.M., Shmerko, V.P., Stanković, R.S., *Decision Diagram Technique for Micro - and Nanoelectronic Design*, CRC Press, Taylor & Francis, Boca Raton, London, New York, 2006. Cited on page(s) 8, 16

[255] Yanushkevich, S.N., Shmerko, V.D., Popel, D., Cheushev, V., Stanković, R.S., "Information theoretic approach to minimization of polynomial expressions over $GF(4)$," *Proc. 30th Int. Symp. on Multiple-Valued Logic*, Portland, Oregon, USA, May 23-25, 2000, 265-270. DOI: 10.1109/ISMVL.2000.848630 Cited on page(s) 49

[256] Zakrevskij, A., "Search Space Reducing: a super - fast algorithm for minimum AND-EXOR implementation of systems of weakly specified Boolean functions," *Proc. Int. Conf. on Pattern Recognition and Information Processing*, Minsk, Belarus, Vol. 1, 1997, 327-331. Cited on page(s) 27

[257] Zakrevskij, A., Jaroszewicz, S., Yanushkevich, S., "Minimization of Reed-Muller expansions for systems of incompletely specified MVL functions," *Proc. Int. Symp. on Methods and Models in Automation and Robotics*, Miedzyzdroje, Poland, Vol. 3, 1996, 1085-1090. Cited on page(s) 27

[258] Zakrevskij, A., Toropov, N., "Optimizing polynomial implementation of incompletely specified Boolean functions," *Proc. Int. Conf. on Computer Aided Design of Discrete Devices*, Minsk, Belarus, Vol. I, 1995, 93-98. Cited on page(s) 48

[259] Zakrevskij, A.D., Zakrevski, L.A., "Fast algorithm for minimizing Reed-Muller expansions of systems of incompletely specified MVL functions," *Proc. 27th International Symposium on Multiple-Valued Logic*, 1997, 61-65. DOI: 10.1109/ISMVL.1997.601375 Cited on page(s) 48

[260] Zhegalkin, I.I., "On the techniques of calculating sentences in symbolic logic," *Math. Sb.*, Vol. 34, 1927, 9-28, (in Russian). Cited on page(s) 15

[261] Zhegalkin, I.I., "Arithmetic representations for symbolic logic," *Math. Sb.*, Vol. 35, 1928, 311-377, (in Russian). Cited on page(s) 15

[262] Zilic, Z., *Towards Spectral Synthesis: Field Expansions for Partial Functions and Logic Modules for FPGAs*, Ph. D. thesis, University of Toronto, Canada, 154 p. Cited on page(s) 48

[263] Zilic, Z., Vranesic, Z., "Current-Mode CMOS Galois Field Circuits," *Proc. 23rd Int. Symp. on Multiple-Valued Logic*, 1993, 245-250. DOI: 10.1109/ISMVL.1993.289552 Cited on page(s) 17, 25

[264] Zilic, Z., Vranesic, Z.G., "A multiple-valued Reed-Muller transform for incompletely specified functions," *IEEE Transactions on Computers*, Vol. 44, No. 8, 1995, 1012-1020. DOI: 10.1109/12.403717 Cited on page(s) 8, 35, 36, 48

[265] Zilic, Z., Vranesic, Z.G., "Reed-Muller forms for incompletely specified functions via sparse polynomial interpolation," *Proc. 25th International Symposium on Multiple-Valued Logic*, 1995, 36-43. DOI: 10.1109/ISMVL.1995.513507 Cited on page(s) 48

[266] Zilic, Z., Vranesic, Z.G., "New interpolation algorithms for multiple-valued Reed-Muller forms," *Proc. 26th Int. Symp. on Multiple-Valued Logic*, 1996, 16-23. Cited on page(s) 48

[267] Zilić, Ž., Vranešić, Z.G., "A deterministic multivariate polynomial interpolation algorithm for small Finite Fields," *IEEE Trans. Computers*, Vol. 37, No. 2, 2002, 1100-1105. DOI: 10.1109/TC.2002.1032628 Cited on page(s) 48

Authors' Biographies

RADOMIR S. STANKOVIĆ

Radomir S. Stanković received a B.Sc. degree in Electronic Engineering from the faculty of Electronics, University of Niš, Serbia, in 1976, and M.Sc. and Ph.D. degrees in Applied Mathematics from the Faculty of Electrical Engineering, University of Belgrade, Serbia, in 1984 and 1986, respectively. Currently, he is a Professor at the Department of Computer Science, Faculty of Electronics, University of Niš, Serbia. In 1997, he was awarded by the Kyushu Institute of Technology Fellowship and worked as a visiting researcher at the Department of Computer Science and Electronics, Kyushu Institute of Technology, Iizuka, Fukuoka, Japan. In 2000 he was awarded by the Nokia Visiting Researcher Fellowship by Nokia, Finland. Since 1999, he has worked in part at the Tampere International Center for Signal Processing, Department of Signal Processing, Faculty of Computing and Electrical Engineering, Tampere University of Technology, Tampere, Finland, where he is currently an adjunct professor. His research interests include switching theory, multiple-valued logic, spectral techniques, and signal processing.

JAAKKO T. ASTOLA

Jaakko T. Astola received B.Sc., M.Sc., Licenciate, and Ph.D. degrees in Mathematics (specializing in error-correcting codes) from Turku University, Finland in 1972, 1973, 1975, and 1978, respectively. From 1976–1977 he was with the Research Institute for Mathematical Sciences of Kyoto University, Kyoto, Japan. Between 1979 and 1987 he was with the Department of Information Technology, Lappeenranta University of Technology, Lappeenranta, Finland, holding various teaching positions in Mathematics, Applied Mathematics, and Computer Science. In 1984, he worked as a visiting scientist at Eindhoven University of Technology, The Netherlands. From 1987–1992 he was an Associate Professor in Applied Mathematics at Tampere University, Tampere, Finland. Since 1993, he has been a Professor of Signal Processing and the Director of Tampere International Center for Signal Processing, leading a group of about 60 scientists and was nominated Academy Professor by Academy of Finland (2001–2006). Currently, he is Director of the Centre of Excellence of the Academy of Finland in Signal Processing. His research interest include signal processing, coding theory, spectral techniques, and statistics. He is a Fellow of IEEE.

CLAUDIO MORAGA

Claudio Moraga received a B.Sc. in Electrical Engineering from the Catholic University of Valparaiso (UCV), Chile, in 1961 and a M.Sc. in Electrical Engineering from the Massachusetts Institute of Technology (MIT), in 1962, and a Ph.D. in Electrical Engineering (Summa cum Laude) from the Technical University "Federico Santa Maria" (UTFSM), Valparaiso, Chile, in 1972. From 1963–1970 he was a lecturer at the Deptartment of Electrical Engineering, UCV, where he served as the head of the department from 1964–1970. From 1970–1973 he worked at the Technical University "Federico Santa Maria" (UTFSM), as an Associate Professor of Electrical Engineering, and was promoted to a Professor of Computer Science in 1973. Also during that time, from 1972–1973, he also served as the Academic Vice President of the Technical University "Frederico Santa Maria" (UTFSM). From November 1974–August 1976, he was awarded the Alexander von Humboldt Research Fellowship at the Department of Computer Science, University of Dortmund, FRG. He continued working at the same University as a lecturer from September 1976–March 1985. From April 1985–September 1986, he was a Professor of Computer Science, in the area of Computer Architecture, at the Department of Mathematics and Computer Science, University of Bremen, FRG. From October 1986–February 2002, he was a Professor of Computer Science, in the area Theory of Automata at the Department of Computer Science, University of Dortmund, FRG. Since March 2002, Claudio Moraga is a Professor Emeritus at the Department of Computer Science, University of Dortmund, FRG. Since March 2006, he has been a Researcher Emeritus at the European Centre for Soft Computing, Mieres, Asturias, Spain. Claudio Moraga is a member of the IEEE Technical Committee on Multiple-valued Logic (USA), and a member of the editorial board of several international journals, as well as holding several awards, including the "Long Service Award for outstanding contributions to Multiple-valued Logic since 1971" from the IEEE TC on Multiple-valued Logic, received in May 2004, and the best paper award in Multiple-valued Logic at the ISMVL 2010. In 2005, Claudio Moraga was awarded the title of the Doctor Honoris Causa of the University of Niš, Serbia.

Index

Printed in the United States
by Baker & Taylor Publisher Services